AF454467

Biomass Wastes for Energy Production

Biomass Wastes for Energy Production

Editor

Eliseu Monteiro

MDPI • Basel • Beijing • Wuhan • Barcelona • Belgrade • Manchester • Tokyo • Cluj • Tianjin

Editor
Eliseu Monteiro
University of Coimbra
Portugal

Editorial Office
MDPI
St. Alban-Anlage 66
4052 Basel, Switzerland

This is a reprint of articles from the Special Issue published online in the open access journal *Energies* (ISSN 1996-1073) (available at: ImprovingMunicipalSolidWasteManagementStrategiesofMontr\unhbox\voidb@x\bgroup\let\unhbox\voidb@x\setbox\@tempboxa\hbox{e\global\mathchardef\accent@spacefactor\spacefactor}\let\begingroup\endgroup\relax\let\ignorespaces\relax\accent19e\egroup\spacefactor\accent@spacefactoral(Canada) UsingLifeCycleAssessmentandOptimizationofTechnologyOptions).

For citation purposes, cite each article independently as indicated on the article page online and as indicated below:

LastName, A.A.; LastName, B.B.; LastName, C.C. Article Title. *Journal Name* **Year**, *Volume Number*, Page Range.

ISBN 978-3-0365-0560-2 (Hbk)
ISBN 978-3-0365-0561-9 (PDF)

Contents

About the Editor

Eliseu Monteiro (Ph.D.) is currently Assistant Professor at the Mechanical Engineering Department, Faculty of Science and Technology of the University of Coimbra. He obtained his Mechanical Engineering degree at UTAD and his Ph.D. in Combustion Sciences at ENSMA of the University of Poitiers. Eliseu Monteiro has over 100 refereed publications, 47 in peer-reviewed Scopus journals, 40 international conferences, 7 book chapters, and 2 books, among others. He has integrated several research teams in 9 national and international research projects. He has served as Guest Editor of 3 Special Issues of the *Energies, Processes,* and *Resources* journals. He is currently co-Guest Editor of the Special Issue of Energies "Renewable Energy from Solid Waste".

Preface to "Biomass Wastes for Energy Production"

This Special Issue of Energies, "Biomass Wastes for Energy Production", comprises 10 papers covering the latest advances on waste-to-energy technologies which contribute to a rethinking of the world's energy supply systems. The environmental problems of the actual world's energy supply systems and the increasing amount of global solid waste production are forcing a fundamental shift toward greater reliance on biomass wastes. Waste-to-energy systems have become a paramount topic for both industry and researchers due to interest in energy production from waste and improved chemical and thermal efficiencies with more cost-effective designs. This biomass shift is also important for industries to become more efficient by using their own wastes to produce their own energy in light of the circular economy concept. For this Special Issue of Energies, we invited submissions exploring the latest advances in the field of waste-to-energy technologies from experimental and computational perspectives. The accepted papers were selected after a rigorous review process demonstrate the highly innovative and informative venue for essential and advanced scientific and engineering research in the field of waste-to-energy technologies. In the first paper, the authors proposed torrefaction processing of various forestry biomass residues as an initial treatment and a means of preparing alternative fuels or substrates for other applications. Based on the results after the torrefaction process, the tested materials were characterized by exceptionally good hydrophobic properties, higher heating value, and higher energy densification. Considering many physical and chemical parameters, such as volatile matter, higher heating value, and fixed carbon content, forestry biomass resembles hard coal after torrefaction but still remains biomass, which is known as an ecological and environmentally friendly source of energy. The second paper studied the possibility of using torrefaction to valorize elephant waste and to determine the impact of technological parameters on the waste conversion rate and fuel properties of resulting biochar. The produced biochars were characterized in terms of moisture content, organic matter, ash, and higher heating values. In addition, thermogravimetric and differential scanning calorimetry analyses were also used for process kinetics assessment. The results show that torrefaction is a feasible method for elephant dung valorization, which could then be used as fuel in households for cooking and heating. Paper three presented a life cycle assessment of the integrated value chain from peach pruning residues for electricity generation as compared with common practice, including the mulching process of the pruned biomass in an orchard. The results show that biomass harvesting, chipping, and its delivery to a power plant is feasible from an environmental point of view. The total global warming potential of this value chain was almost 12 times lower than the mulching and leaving of the pruned biomass in an orchard. In the fourth paper, the authors carried out a life cycle assessment for the current and proposed waste management system in Montréal city. Using life cycle assessment results, a non-dominated sorting genetic algorithm was used to optimize the waste flows. The optimization showed that the current recovery ratio of organic waste of 23% could be increased to 100% recovery of food waste. Moreover, recycling could be doubled, and landfilling halved. By using a three-objective optimization algorithm, the optimized waste flow for Montréal results in 2% of waste to anaerobic digestion, 7% to compost, 32% to recycling, 1% to incineration, and 58% to landfill. The fifth paper studied the performance evaluation of anaerobic digestion of dairy wastewater in a multi-section horizontal flow reactor equipped with microwave and ultrasonic generators to stimulate biochemical processes. They found that organic loading rates had the greatest impact on the effects of anaerobic digestion of dairy wastewater in terms of organic compound removal and biogas and methane yields. They also found a significant impact of ultrasonication on the effects of anaerobic digestion of dairy wastewater. Increase in ultrasonic intensity significantly reduced the efficiency of organic compounds' removal from wastewater, as well as biogas yield. Paper six provides an overview of biomass streams that can be used for biogas production and their alternative uses. Their literature review was performed using the machine learning technique "co-occurrence analysis of terms". They conclude that a large share of the biomass streams considered in the biogas estimates have many alternative uses, which limit their contribution to future biogas production, and there are streams not being considered in estimates for biogas production although they have the proper characteristics. The seventh paper determined generalized energy and ecological characteristics of steam boilers co-firing hard coal with biomass. The energy characteristics determined are the dependence of the gross energy efficiency of boilers on such decision parameters as their efficiency and the share of biomass chemical energy in fuel. The ecological characteristics are the dependence of gaseous emission streams and dust on the same decision parameters. Boiler characteristics can be used when forecasting the impact of changes in operating conditions on the effects achieved in existing, modernized, and designed boilers. Paper eight reported the main results of the experimental activity to optimize and develop a fixed-bed updraft gasification process for power generation from biomass. Particular attention was paid to the optimization of an integrated double stage wastewater management system designed to minimize both liquid residues and water content. They identified the optimal process parameters for the operation of the syngas cleaning section that resulted in a 60% reduction of wastewater disposal. In the ninth paper, the authors performed experimental and modeling analysis of brewers' spent grains gasification in a pilot-scale downdraft reactor. A ratio of 1.3 kg of brewers' spent grains per kWh of electricity generated was obtained, with an average electrical efficiency of 16.5%. A modified thermodynamic equilibrium model of the downdraft gasification is developed to assess the potential applications of the main Portuguese biomasses through produced gas quality indices. They conclude that in using air as a gasifying agent, biomass gasification provides produced gas of sufficient quality for use in energy production in boilers or turbines. The last paper provided a comparative technoeconomic analysis of small-scale gasification systems for electricity generation in a 15 kWe downdraft gasifier. A spreadsheet economic model was developed combining the net present value, internal rate of return, and the payback period. A Monte Carlo sensitivity analysis was used to measure the performance of the economic model and determine the investment risk. The analysis showed an electricity production between 11.6 to 15 kW, with a general system efficiency of approximately 13.5%. The viability of the projects was predicted for an internal rate of return between 16.88 to 20.09% and a payback period between 8.67 to 12.61 years. This study highlights the empowering effect of small-scale gasification systems settled in decentralized communities for electric power generation. To conclude, special gratitude and appreciation is extended to all the authors for their high-quality submissions and the anonymous reviewers for volunteering their time and expertise to evaluate the scientific merit of the submitted papers. Additionally, the Special Issue assistant editor, Mr. Allen Liu, deserves special thanks for his great effort and support in making this Special Issue successful. With this Special Issue of Energies, we hope to highlight new contributions in the growing and stimulating interdisciplinary field of waste-to-energy technologies.

Eliseu Monteiro
Editor

Article

Alternative Fuels from Forestry Biomass Residue: Torrefaction Process of Horse Chestnuts, Oak Acorns, and Spruce Cones

Arkadiusz Dyjakon * and Tomasz Noszczyk *

Institute of Agricultural Engineering, Wroclaw University of Environmental and Life Sciences,
51-630 Wroclaw, Poland
* Correspondence: arkadiusz.dyjakon@upwr.edu.pl (A.D.); tomasz.noszczyk@upwr.edu.pl (T.N.);
  Tel.: +48-71-320-5945 (A.D.)

Received: 19 April 2020; Accepted: 12 May 2020; Published: 14 May 2020

Abstract: The global energy system needs new, environmentally friendly, alternative fuels. Biomass is a good source of energy with global potential. Forestry biomass (especially wood, bark, or trees fruit) can be used in the energy process. However, the direct use of raw biomass in the combustion process (heating or electricity generation) is not recommended due to its unstable and low energetic properties. Raw biomass is characterized by high moisture content, low heating value, and hydrophilic propensities. The initial thermal processing and valorization of biomass improves its properties. One of these processes is torrefaction. In this study, forestry biomass residues such as horse chestnuts, oak acorns, and spruce cones were investigated. The torrefaction process was carried out in temperatures ranging from 200 °C to 320 °C in a non-oxidative atmosphere. The raw and torrefied materials were subjected to a wide range of tests including proximate analysis, fixed carbon content, hydrophobicity, density, and energy yield. The analyses indicated that the torrefaction process improves the fuel properties of horse chestnuts, oak acorns, and spruce cones. The properties of torrefied biomass at 320 °C were very similar to hard coal. In the case of horse chestnuts, an increase in fixed carbon content from 18.1% to 44.7%, and a decrease in volatiles from 82.9% to 59.8% were determined. Additionally, torrefied materials were characterized by their hydrophobic properties. In terms of energy yield, the highest value was achieved for oak acorns torrefied at 280 °C and amounted to 1.25. Moreover, higher heating value for the investigated forestry fruit residues ranged from 24.5 MJ·kg^{-1} to almost 27.0 MJ·kg^{-1} (at a torrefaction temperature of 320 °C).

Keywords: biomass residues; forestry; torrefaction; thermal treatment; biomass valorization; torrefied material properties

---

1. Introduction

Over the past several years, global energy demand has increased significantly. This is associated with economic and industrial development in many countries and energy (heat and electricity) is an essential service required by people all over the world. The total energy consumption in the world in 2017 was 23,696 TWh, which was an increase of about 117.4% compared to 1990 [1]. Unfortunately, the share of coal in total energy generation is high (41.6% in 2017) and its consumption is still growing [1]. In the last few decades, next to energy security, protection of the natural environment has also become a very important issue. The world is now focused on global warming problems and the uncontrolled increase in the global temperature that could lead to ecological disaster [2–4]. Therefore, many measures need to be taken to prevent global warming of more than 1.5 °C [4].

One of the solutions is to further increase the use of renewable energy sources (RES) in total energy production, including biomass residues. In 2019, energy potential from residual biomass in the EU-27 amounted to 8500 PJ·y^{-1}, of which 3800 PJ·y^{-1} was from straw and 3200 PJ·y^{-1} was from forestry residues [5]. It is estimated that in 2050 the global technical potential of biomass residues (from agriculture, forestry, dungs, and organic waste) will range from 55 to 325 EJ·y^{-1} [6]. Moreover, the EU countries have included a share of biomass feedstock in their energy mixes in the strategies formulated for 2030 and 2050 [7]. However, the focus is on the identification and utilization of new alternative sources of biomass, especially residual sources. A circular economy, in particular, the circular waste management sector requires three important factors: financial stability, environmental friendliness, and social wellbeing. Biomass valorization methods, e.g., torrefaction of forestry waste materials adds new value to the forest. Furthermore, thanks to the torrefaction process of waste, the forest biomass can be used to complement many natural products besides bioenergy. It allows for energy diversification by the manufacturer and increases the chance that the cost of bioenergy will be sustainable, while potentially decreasing costs in the energy supply chain [8]. Thus, valorization of biomass residues is recommended to improve their properties and to expand their application options [9]. This is also a result of the fact that direct use of biomass is associated with several difficulties. Raw, untreated waste biomass is characterized by a heterogeneous structure, higher moisture absorption capabilities (hydrophilic properties), and significantly lower heating value [10,11]. In addition, these negative properties decrease the economical use of raw biomass in practice. Therefore, it is necessary to use various pre-treatment processes (Figure 1). The use of mechanical, thermal, or biological treatments significantly improves the physical properties of raw biomass [9,12,13]. It should also be noted, that forestry biomass residues can support the carbon sequestration process. Loehle [14] analyzed the sequestration of carbon by commercial forestry. It was observed, that biomass used for energy production has a value of 100 years of sequestration, which corresponds to twelve tons of avoided carbon emission. Thus, the utilization of forestry waste biomass is also a solution to climate change risks.

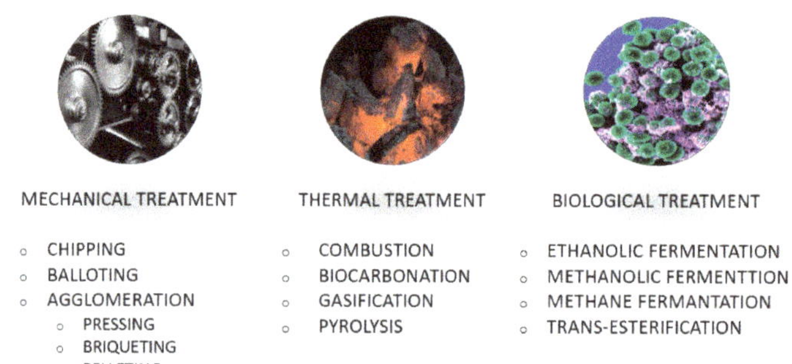

Figure 1. Pre-treatment methods of raw biomass.

One of the most promising and commonly used methods of converting biomass is thermal pre-treatment [15,16]. Among the thermal processing of biomass, torrefaction has become very attractive [17]. Torrefaction is also known as low-temperature pyrolysis (usually the pyrolysis process is carried out in temperatures up to 600–800 °C [18,19]) with high-temperature drying or roasting. The torrefaction process relies on heating biomass in atmospheric pressure and inert conditions. The typical range of temperature is from 200 °C to 320 °C [10,20,21]. To maintain the inert atmosphere in the heating chamber, carbon dioxide, or nitrogen flow is applied inside the reactor [22,23]. The non-oxygen atmosphere is required to prevent the combustion of biomass [24]. Therefore, thermal processing of biomass in a non-oxygen atmosphere (especially lignocellulosic biomass) allows for the decomposition of lignin, cellulose, and hemicellulose, which is important for further processing of biomass [9]. The depolymerization of the lignin, cellulose, and hemicellulose depends on the temperature and the duration of the process. The hemicellulose fibers are the first to depolymerize, due to its low molecular weight, and this is followed by lignin. The hardest fiber to depolymerize is cellulose and it is the last fiber to undergo depolymerization [25–27]. The content of these three basic fibers in biomass has an effect on its degree of degradation and its activity [28]. After the torrefaction process, torrefied biomass is characterized by a higher energy densification ratio (EDR) in comparison to the raw biomass. The energy densification increases by approximately 30%. This is because the loss of mass is larger (approximately 30–35%, depending on the material and the temperature of the torrefaction process) than the loss of energy (10–15%) [28–30]. The loss of mass is associated mainly with moisture removal from biomass and the initial devolatilization (thermal decomposition). Lower and limited moisture content in torrefied biomass (1–3%) affects the water gas shift reaction and the increase in the hydrogen content in syngas [31,32]. Torrefied biomass is advantageous because of its variable bulk density as well as its higher heating value (HHV), which results in savings in transport, storage, and further processing [10,33,34]. Torrefaction also results in better hydrophobic properties in biomass [35] due to the loss of hydroxyl groups during the process. This allows torrefied biomass to be stored in the open space for a long period, with a low risk of dampness, decomposition, and decay [10,33,34] in comparison to raw biomass (which tends to decompose quickly and is sensitive to external weather conditions [36,37]). Finally, torrefied biomass has better grindability due to the decomposition of fibers during thermal processing [33,35].

However, the process temperature has an influence on the costs, which should be economically justified from a practical point of view. The higher temperature of the torrefaction process results in improvement in the physical properties of the final product from biomass, but it also raises the energy inputs (and costs) to produce this alternative fuel. Energy consumption of a muffle furnace (in laboratory research) for a torrefaction temperature of 300 °C was found the be more than 60% higher compared to a temperature of 200 °C (with the same residence time of 60 min) [10]. Thus, the proper selection of the torrefaction temperature in terms of the expected valorized biomass properties is a very important factor in the industrial planning of the thermal processing of raw biomass.

Therefore, knowledge about the properties of the torrefied biomass residues and the changes caused by the temperature of the thermal process are crucial with regard to further application options, process modeling, and its economical use in practice. No data about the torrefaction process of the fruits of deciduous and coniferous trees was found in the literature.

This work aimed to assess the effect of the torrefaction process on the selected physical and chemical properties of investigated biomass residues from the forestry sector, such as horse chestnuts, oak acorns, and spruce cones. Specifically, the study investigated the influence of the temperature of the torrefaction on (i) changes in the results of the proximate analysis of torrefied materials, (ii) the hydrophobic propensities of final products, and (iii) changes in the basic physical parameters of the obtained products.

2. Materials and Methods

2.1. Materials Used in the Research

The subject of the research was different types of fruit from forest trees such as chestnut, oak, and spruce. In detail, the following three types of fruit were investigated (Figure 2): (a) horse chestnuts, (b) oak acorns, and (c) spruce cones.

(a)   (b)   (c)

Figure 2. Organic materials used in the studies: (**a**) horse chestnuts; (**b**) oak acorns; and (**c**) spruce cones.

2.2. Samples Preparation and Torrefaction Procedure

The preparation of the materials included their initial drying in the chamber KBC–65 W (WAMED, Warszawa, Poland) (Figure 3a) to obtain an analytical state for all samples. The drying temperature and duration time was 105 °C and 24 h, respectively. Next, horse chestnuts, oak acorns, and spruce cones were comminuted in the mill LMN 400 (TESTCHEM, Pszów, Poland) (Figure 3b). The size of the sieve was 1 mm. Finally, the samples (50 g each) were torrefied in the electrically-heated muffle furnace SNOL 8.2/1100 (SNOL, Utena, Lithuania) (Figure 3c). The mass of the samples, before and after the torrefaction process was measured using the scale RADWAG AS 220.R2 (RADWAG, Radom, Poland) (Figure 3d).

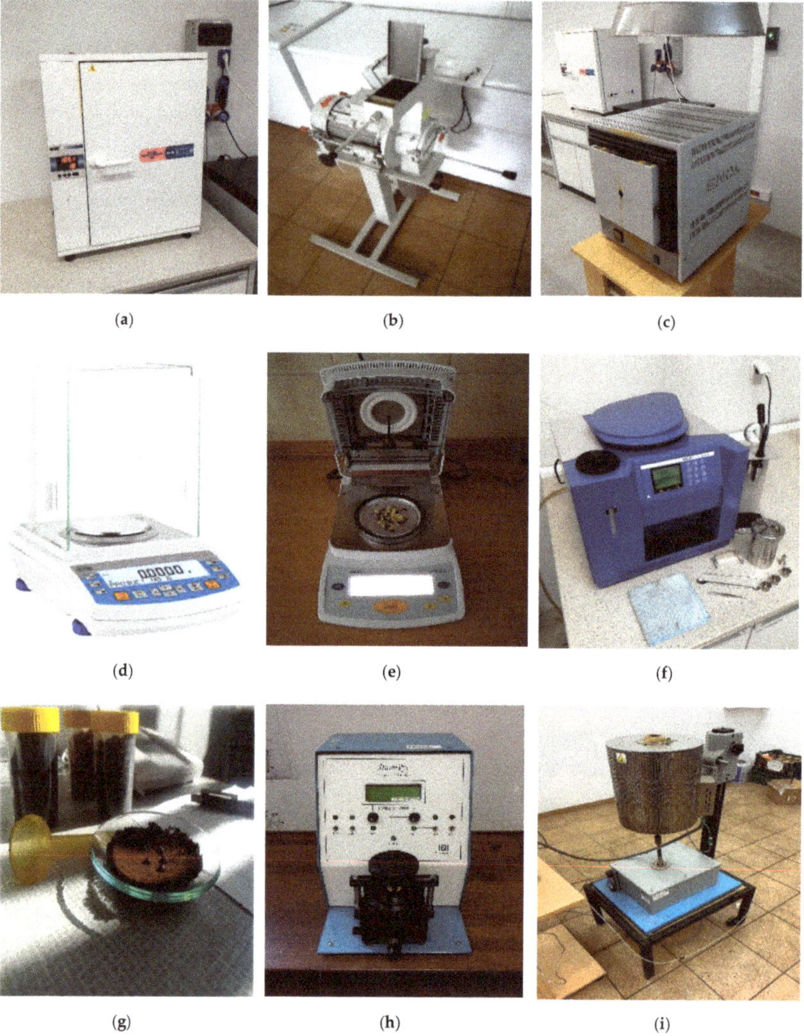

Figure 3. Laboratory devices: (**a**) drying chamber KBC-65W; (**b**) biomass mill LMN 400; (**c**) muffle furnace SNOL82/1100; (**d**) scale RADWAG AS 220.R2; (**e**) moisture analyzer SARTORIUS MA150; (**f**) calorimetric bomb IKA C200; (**g**) a set for WDPT (water drop penetration time) test; (**h**) gas pycnometer HumicPyc; and (**i**) thermogravimetric analyzer.

The torrefaction temperatures were 200 °C, 220 °C, 240 °C, 260 °C, 280 °C, 300 °C and 320 °C. Carbon dioxide (to maintain the inert atmosphere) from the gas cylinder was used in the reactor chamber (90 mL·min^{-1}). The duration time of the torrefaction process was 60 min. The proposed thermal processing of forestry biomass, its basic properties, and process conditions are shown in Figure 4.

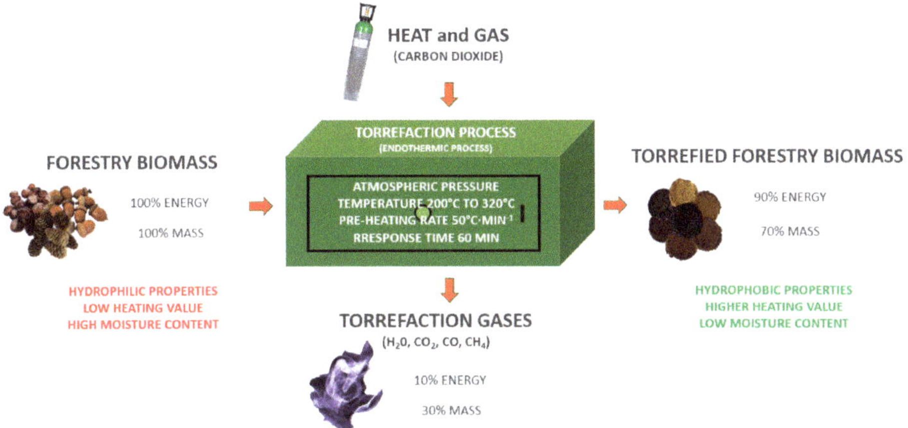

Figure 4. Forestry biomass (horse chestnuts, oak acorns, spruce cones) processing in the torrefaction process.

After the torrefaction process and cooling of the chamber, the samples were closed in an airtight plastic bag (to protect the material from absorbing water from the air). The process was repeated three times.

2.3. Proximate Analysis

The physical properties of the torrefied forestry biomass residues were determined to perform the proximate analysis. All the samples were taken following the applied ISO Standards. The following parameters were determined: ash content (AC), moisture content (MC), volatile matter content (VMC), higher heating value (HHV), and fixed carbon content (FCC). The measures were repeated five times.

The moisture content was determined using a laboratory moisture analyzer SARTORIUS MA150 (Sartorius, Goettingen, Germany) (Figure 3e). In this test, the PN-EN ISO 18134-2:2017-03E standard was applied [38]).

Ash content in the raw and torrefied forestry biomass was determined according to PN EN ISO 18122:2015 [39] using the muffle furnace SNOL 8.2/1100 (SNOL, Utena, Lithuania) and the following formula:

$$AC = \frac{m_A - m_C}{m_M - m_C} \cdot 100\% \tag{1}$$

where AC is the ash content in the material (%), m_A is the a mass of the crucible with ash after heating (g), m_C is the mass of the empty crucible (g), and m_M is the mass of the crucible with the material before heating (g).

The volatile matter content (VMC) in the investigated materials was determined by applying PN-EN ISO 18123:2016-01 standard [40]. The formula is as follow:

$$VMC = \frac{1 - (m_S - m_C)}{m_M} \cdot 100\% \tag{2}$$

where *VMC* is the content of volatile matter in the material (%), m_S is the mass of the crucible with fuel sample after heating (g), m_C is the mass of the empty crucible (g), and m_M is the mass of the crucible with fuel sample before heating (g).

To determine the higher heating value (HHV) of the material a calorimetric bomb IKA C200 (IKA, Lucknow, India) (Figure 3f) was used. The measurement was performed following PN-EN ISO 18125:2017-07 [41].

The fixed carbon content (the solid combustible residue) in the raw and torrefied forestry biomass was determined following ASTM D-3172-73 [42] using the following formula:

$$FCC = (1 - MC - AC - VMC) \cdot 100\%$$

(3)

where *FCC* is the fixed carbon content in raw and torrefied forestry biomass (%), *MC* is the moisture content in raw and torrefied forestry biomass (%), *AC* is the ash content in raw and torrefied forestry biomass (%); and VMC is the volatile matter content in raw and torrefied forestry biomass (%).

2.4. Additional Analysis

The additional properties of the alternative fuel from horse chestnuts, oak acorns, and spruce cones were also investigated by performing hydrophobicity tests, as well as the measurement of bulk density (ρ_B), specific density (ρ_S), and porosity (ε). Mass yield (MY), the energy densification ratio (EDR), and energy yield (EY) were determined based on the physical properties. The investigations were repeated five times.

The hydrophobic properties were determined by the water drop penetration time (WDPT) test [43]. Raw and torrefied material weighing 5 g was spread (in a thin layer of 2 mm) on a laboratory slide glass (Figure 3g). Next, the five drops of distilled water were put on the surface of the material. The temperature of the distilled water was 20 °C. Then, the time it took the water drop to the penetrate through the layer of the investigated material was measured with a stopwatch. The hydrophobic properties were determined by comparing the values obtained for the drop penetration time to the classification data (Table 1).

Table 1. Classification criterion of hydrophobic properties [10,44].

Classification Criterion Time of the Penetration of a Drop of Water	Hydrophobic Properties
<5 s	Hydrophilic
5–60 s	Slightly hydrophobic
60–600 s	Strongly hydrophobic
600–3600 s	Severely hydrophobic
>3600 s	Extremely hydrophobic

In the case where the penetration time of distilled water drop exceeded 1 h, the sample was covered with the lids (to minimize the influence of the evaporation process) [43].

The bulk density of the material was determined according to PN-EN 1237:2000 standards [45] using the following formula:

$$\rho_B = \frac{m_i}{V_i} \cdot 100\%$$

(4)

where ρ_B is the bulk density (kg·m^{-3}), m_i is the mass of the material in the container (kg), and V_i is the volume of the container (m^3).

The specific density of raw and torrefied material (ρ_S) was determined using the gas pycnometer HumicPyc (InstruQuest Inc., Boca Raton, USA) (Figure 3h) whereas, the porosity was calculated according to PN-EN 1936:2010 [46] using the following formula:

$$\varepsilon = \left(1 - \frac{\rho_S}{\rho_B}\right) \cdot 100\% \tag{5}$$

where ε is the porosity of the material in the dry analytical state (%), ρ_B is the bulk density of the material in the dry analytical state (kg·m^{-3}), and ρ_S is the specific density of material in the dry analytical state (kg·m^{-3}).

Mass yield (MY) (using the thermogravimetric analyzer (Figure 3i), energy densification ratio (EDR) and energy yield (EY) was determined using the following formulas [28,29,47]:

$$MY = \frac{m_C}{m_R} \cdot 100\% \tag{6}$$

where MY is the mass yield (%), m_C is the mass of dry torrefied material (g), and m_R is the mass of raw material (g).

$$EDR = \frac{HHV_C}{HHV_R} \tag{7}$$

where EDR is the energy densification ratio (-), HHV_C is the higher heating value of torrefied material (kJ·kg^{-1}), and HHV_R is the higher heating value of raw material (kJ·kg^{-1}).

$$EY = MY \cdot EDR \tag{8}$$

where EY is the energy yield (%), MY is the mass yield (%), and EDR is the energy densification ratio (-).

The results for the proximate analysis, the additional properties, and diagrams were developed in statistical software STATISTICA (StatSoft-DELL Software, Texas, USA). The detailed results, including standard deviations and coefficient of variation are available in the Supplementary Materials, Tables S1–S5.

3. Results and Discussion

The first noticeable change in properties after torrefaction was the change in the color of the torrefied materials (Figure 5). The raw materials were characterized by a bright brown color. However, as the torrefaction process temperature increased, the color of the materials got darker (from brown to dark brown to black). Horse chestnuts, oak acorns, and spruce cones torrefied at 300 °C and higher were black in color and looked like fine coal. The change in color is a natural phenomenon during torrefaction or roasting of the biomass materials. The occurrence of a similar change in color was also observed by other researchers [48,49].

3.1. Results of the Proximate Analysis

The investigated raw materials were characterized by a moisture content of 30.5% for the horse chestnuts, 30.6% for the oak acorns, and 10.2% for the spruce cones. After the torrefaction process, the ash content in the tested materials increased as the temperature of the process increased (Figure 6). For the chestnuts, the ash content was in the range from 2.3% (for raw material) to 6.1% (for torrefied material 300/320 °C). For oak acorns and spruce cones, it ranged from 2.1% to 4.4% and from 0.9% to 2.4%, respectively. The results showed that there was up to a three-fold increase in ash content in the torrefied material. Such a significant increase in ash content in the final product is primarily due to the weight loss in the form of volatile matters released during the thermal decomposition under torrefaction conditions. Moreover, the higher the process temperature, the greater the volatile matter loss. As a result, the percentage share of non-combustible substances (ash) remaining in the material increases. An increase in ash content in the material after the torrefaction process has also been observed in other research, where the AC was higher for material torrefied at 300 °C in comparison to material torrefied at 200 °C [10,16,50]. However, the increase of the ash content after torrefaction is not a significant disadvantage of this process. For comparison, the AC in coal is ca. 25–35% [51], so an AC of 4% or 6% in torrefied biomass residues is still competitive compared to coal. Alternative fuels from typical untreated wooden/agricultural biomass in the form of a pellet, also had an ash content in the range from 0.66% to 9.43%, depending on the type of material [52]. Interestingly, despite the increased ash content, the torrefied forest residues were still within the range of values found for thermally unprocessed biomass materials. This is a positive feature in the context of the materials studied in this work.

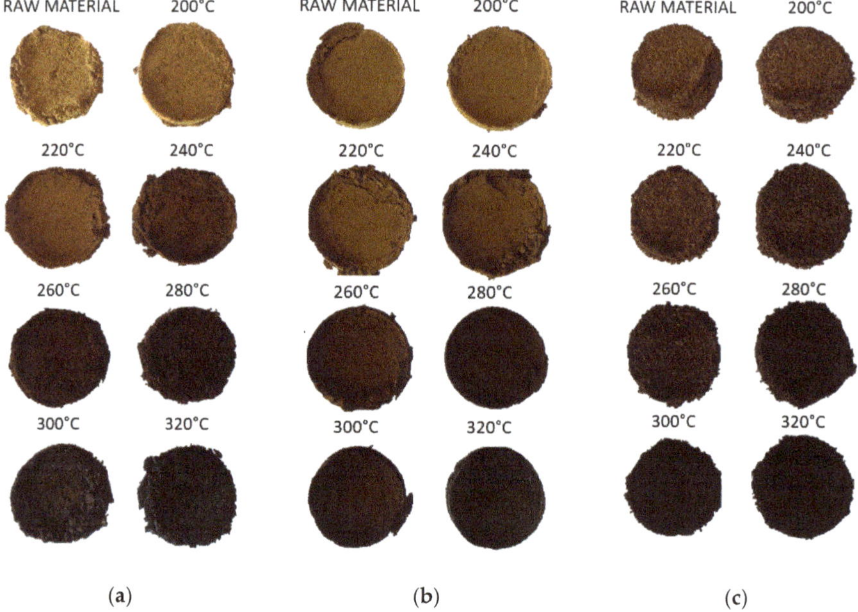

Figure 5. Color change in the material depending on the torrefaction process temperature: (**a**) horse chestnuts; (**b**) oak acorns; and (**c**) spruce cones.

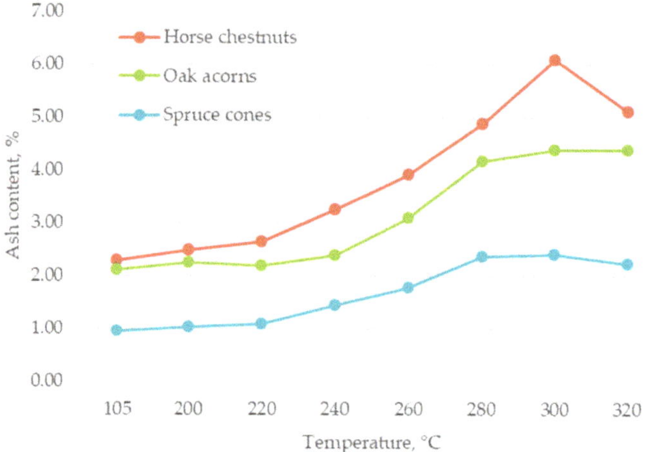

Figure 6. Ash content in the torrefied horse chestnuts, oak acorns, and spruce cones.

An interesting phenomenon was observed in horse chestnuts torrified at a higher temperature (320 °C). At this temperature, the ash content decreased remarkably. This trend can also be seen in the case of oak acorns and spruce cones, although it is not as obvious. This is probably related to the vaporization of alkali salts [53], mainly alkali halides of metals. At a certain temperature of pyrolysis, these compounds move from a solid phase to the gas phase, thus the ash content may decrease [54]. It may also be related to the inhomogeneous structure of materials [55]. This study used whole materials, both shell and pulp, which can be characterized by different content of ash.

The volatile matter content in fuel (fossil or renewable) is directly related to the ease of ignition of the fuel and its combustion stability. Volatile matter is the combustible part of the fuel. So, a material characterized by a high content of volatile matter is easier to ignite and requires less energy for this process. However, fuels with higher VMC emit less heat because a significant part of the mass is released with volatiles to the atmosphere, and the coke (char) residue is the main source of heat/energy [56].

All of the tested forestry biomasses were characterized by decreasing volatile matter content after the torrefaction process; the VMC change depended on the process temperature (Figure 7). The content of volatiles in horse chestnuts was from 79.6% for raw biomass to 50.2% for material torrefied at 320 °C. The VMC in oak acorns went from 82.9% in the raw material to 59.8% in the material torrefied at 320 °C. The oak acorns had the highest content of volatiles in both the raw and torrefied material. In the case of spruce cones, the VMC decreased from 70.9% to 55.3%.

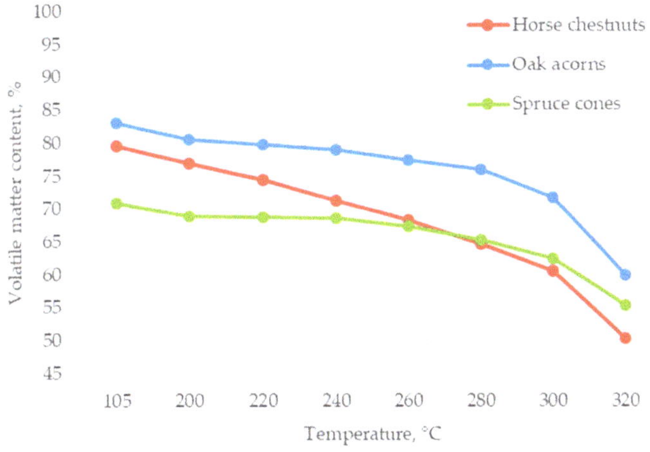

Figure 7. Volatile matter content in the torrefied horse chestnuts, oak acorns, and spruce cones.

The obtained results and behavior trends are similar to those found by other researchers. Tong et al. [57] determined the volatile matter content in raw biomass as being 75–85%. This is an average value of the VMC in raw biomass (i.e., wood, straw, and forestry residues), and is in line with the VMC in horse chestnuts, oak acorns, and spruce cones. Additionally, some similarities can be seen in comparing the volatile matter content in torrefied horse chestnuts, oak acorns, and spruce cones. As the temperature increased, the VMC decreased. Biomass like rice straw, pine sawdust, or other straws are also characterized by volatiles at 40–50% after torrefaction at 300 °C [57,58]. The mean content of volatiles in coal is 40% [56,59], so, torrefied forestry biomass starts to be very close to coal in terms of volatile matter. However, volatiles are also related to volatile organic compounds (VOCs). Biomass is characterized by the different content of VOCs. In biomass like wood or straw, over one hundred types of VOCs have been identified that are toxic and polluting air [60]. This is important for the forest product industry or bioenergy sector because it limits the biomass application options due to environmental regulations.

Fixed carbon is a solid combustible residue in the fuel. With a significant decrease in volatiles and a slight increase in ash content, the total content of fixed carbon should increase in torrefied biomass as the temperature rises. This is because the amount of lost carbon contained in volatile compounds (i.e., light hydrocarbons) is low. This dependence was observed in the obtained results (Figure 8). Fixed carbon content ranged from 18.1% to 44.7%, from 14.9% to 35.5%, and from 28.1% to 42.5%, for horse chestnuts, oak acorns, and spruce cones, respectively. A higher content of FCC is beneficial as it allows more heat to be generated from the material. Correia et al. [58] studied several biomass materials (including straw) during torrefaction and obtained fixed carbon contents ranging from 9.9% to 17.3% for raw biomass, and its content was higher when the torrefaction temperature was higher. Similar amounts of FCC in the biomass before and after torrefaction was noted in other studies [10,61]. As a comparison, the mean fixed carbon content in coal is ca. 60% (dry ash-free state) [56].

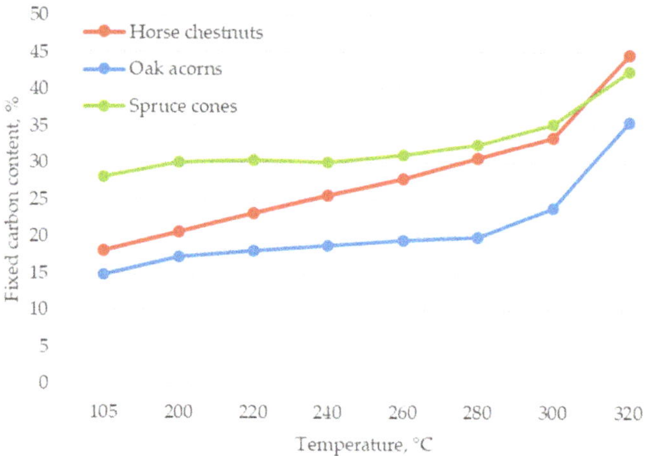

Figure 8. Fixed carbon content in the torrefied horse chestnuts, oak acorns, and spruce cones.

The higher heating value provides basic information about a fuel and its properties. If the HHV is higher, more energy can be obtained from the material during its combustion. The higher heating values for the dry horse chestnuts, oak acorns, and spruce cones ranged from 18.2 MJ·kg^{-1} to 19.6 MJ·kg^{-1} (Figure 9), which is comparable to other types of biomass, like straw or sawdust [52,57,62]. For the forestry biomass, an increase in the HHV due to thermal conversion was observed. Potential deviations from the upward trend probably result from the heterogeneous structure/composition of the investigated material. For example, the horse chestnut possesses a shell and pulp, whose share in the material is not the same. Besides, the measurement error associated with the method of HHV determination can also slightly influence the final result. Horse chestnuts were characterized by an HHV = 26.5 MJ·kg^{-1} after torrefaction at 320 °C. Oak acorns and spruce cones, at the same temperature, reached 25.6 MJ·kg^{-1} and 24.7 MJ·kg^{-1}, respectively. The values of the HHV were much higher after torrefaction than in the raw material. The impact of torrefaction on the higher heating value of biomass has also been confirmed by other researchers [10,57,58,63] who found an increase in the HHV from 18.2 MJ·kg^{-1} to 28.5 MJ·kg^{-1} for waste biomass [10], an increase from 15.9 MJ·kg^{-1} to 21.4 MJ·kg^{-1} for straw [57,58], and from 18.2 MJ·kg^{-1} to 23.2 MJ·kg^{-1} for wooden sawdust [63].

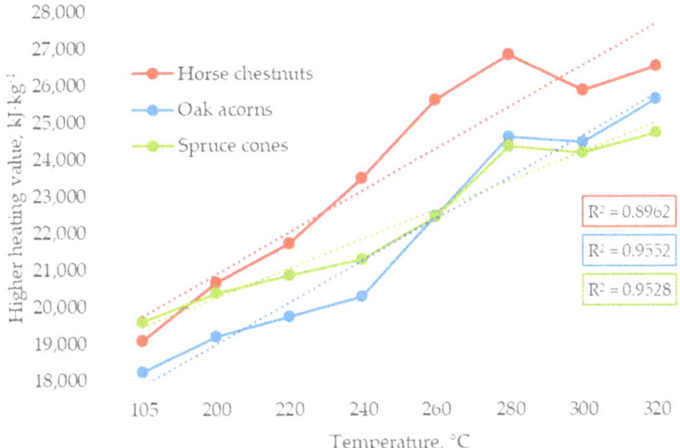

Figure 9. Higher heating value of the torrefied horse chestnuts, oak acorns, and spruce cones.

3.2. Results of Additional Analysis

Beyond the proximate analysis, which is required for the fuels, additional analysis can also provide valuable information about the properties of new, alternative fuels. One of these analyses is the hydrophobicity test. It was observed that the torrefaction process results in the biomass obtaining better hydrophobicity properties compared to the raw material. Thermal processing of biomass allows the decomposition of hydroxyl groups, which are responsible for binding the moisture in the material [64,65]. Thanks to this, biomass after pyrolysis at a lower temperature is characterized by hydrophobic properties, in contrast to the hydrophilic properties of the raw material. The graphical illustration of these properties is shown in Figure 10. It can be observed that drops of distilled water sink into raw material (Figure 10a), but persist on torrefied biomass at 300 °C (Figure 10b).

(a)   (b)

Figure 10. Type of hydrophobic properties of torrefied biomass: (**a**) hydrophilic (105°C); (**b**) extremely hydrophobic (300 °C).

In accordance with the water drop penetration time test for determining hydrophobic properties, material that reaches the value over the 3600 s time period is recognized as extremely hydrophobic material (Table 1). As can be seen in Figure 11, horse chestnuts, oak acorns, and spruce cones obtained extremely hydrophobic properties at different temperatures of torrefaction. For the horse chestnuts, it was 260 °C, and for the oak acorns it was 220 °C. The obtained results are satisfying because hydrophobic properties were achieved at low torrefaction temperatures. Food waste biomass like nuts shells or fruit peels and seeds are characterized by extremely hydrophobic properties at a similar range of temperature. Orange peels are extremely hydrophobic at 240 °C, walnut shells at 220 °C, and pumpkin seeds at 260 °C [10]. The favorable effect of torrefaction on the hydrophobicity of the biomass was also noticed by Alvarez et al. [66] and Pouzet et al. [67]. However, a very interesting result was obtained for spruce cones, which were already extremely hydrophobic in the untreated state (raw material). This probably arises from the internal multilayer structure of the spruce cones and the content of resin. There are studies about the content of resin acids in dependence on the type of tree. Coniferous trees (e.g., pine or spruce) are characterized by a higher content of resin acids than deciduous trees (i.e., beech or oak) [68]. The specific content of resin acids in different parts of the tree was also determined by Eberhardt et al. [69]. This is important from a practical point of view because it means that they can be stored in the open air without any cover or roofing. The hydrophobic properties of torrefied biomass can be used in furniture production (garden, kitchen, bathroom) where water-resistance of the final product is crucial. However, it should be checked that the thermal processing of the material does not adversely affect its other material properties.

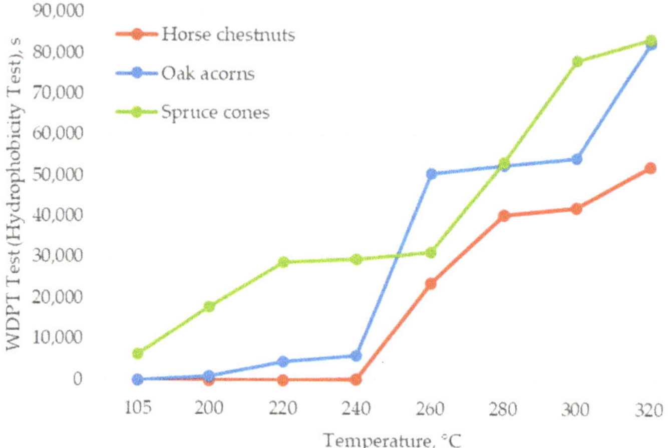

Figure 11. Hydrophobic properties of the raw and torrefied forestry biomass.

Considering the bulk density (ρ_B) of the forestry biomass, it can be seen that it decreases as the temperature of torrefaction increases (Figure 12). The bulk density of horse chestnuts ranged from 0.581 g·cm^{-3} (raw material) to 0.365 g·cm^{-3} (320 °C). For the oak acorns, the ρ_B was from 0.544 g·cm^{-3} to 0.369 g·cm^{-3} and for the spruce cones it ranged from 0.483 g·cm^{-3} to 0.327 g·cm^{-3}, for the raw material and torrefied at 320 °C, respectively. A slight increase in the specific density (ρ_S) of the tested materials was noticed after the torrefaction (Figure 12). Forestry biomass residues torrefied between 200 °C and 300 °C did not show a change in the specific density. The specific density only increased significantly at 320 °C. This may be associated with a further release of bound moisture and increased thermal decomposition of the substance. This translates into an increase in pores in torrefied products and further concentration of heavier compounds in the specific volume of the material. As can be seen in Figure 13, torrefaction causes an increase in the material's porosity. The porosity of the raw material was 59.6% (horse chestnuts), 66.1% (oak acorns) and 68.0% (spruce cones). With increasing temperature, the porosity reached 81.2%, 83.8%, and 85.1%, respectively. The highest increase in porosity was observed between 300 °C and 320 °C. The relationship between bulk density, porosity, and the torrefaction process has also been described by Nhuchhen et al. [70] and Bach et al. [71]. In their studies, torrefaction caused a decrease in the bulk density of biomass, while the porosity index tended to grow.

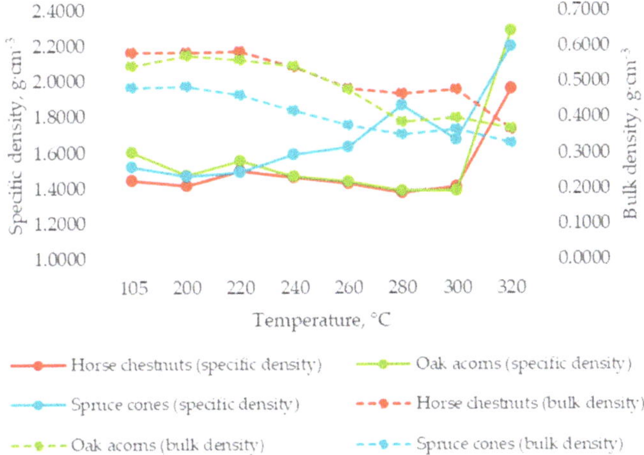

Figure 12. Specific and bulk density of the raw and torrefied forestry biomass.

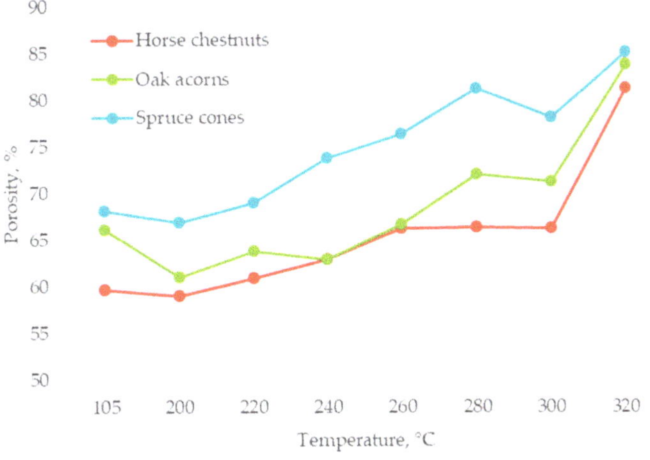

Figure 13. Porosity of the raw and torrefied forestry biomass.

As is well-known, thermal processing causes a mass loss of the treated material. This is associated with both moisture evaporation and the release of volatile substances from the material. Mass yield (mass remaining after torrefaction) depends on the moisture, ash, carbon, and volatile matter content in the biomass. In Figure 14, the mass yield of the tested forestry biomass residues is shown. The highest mass yield (after torrefaction at 320 °C) was 70.6% (mass loss 29.4%), which was observed for the horse chestnuts. For the oak acorns the mass yield was 76.8% (mass loss 23.2%) and for spruce cones it was 84.4% (mass loss 15.6%). A similar process was also observed in other studies. The mass loss for torrefied wood at 280 °C was ca. 37%, and was higher than for horse chestnuts, oak acorns, and spruce cones [72]. The higher mass loss at higher temperatures was also observed for fruit residues [55]. For torrefied oil palm empty fruit bunches at 300 °C, the mass yield amounted to 56% (mass loss at 44%).

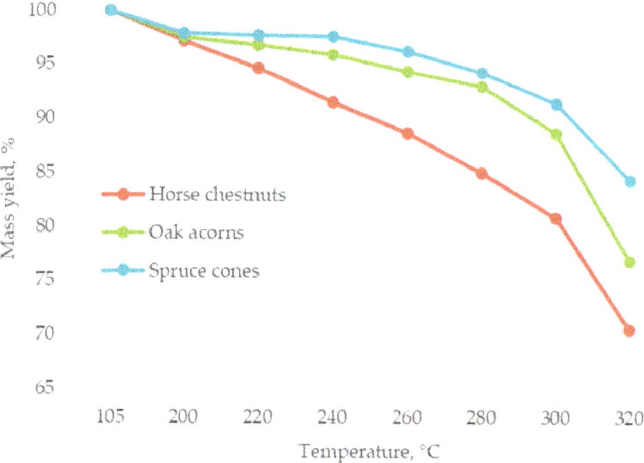

Figure 14. Mass yield of the torrefied forestry biomass.

The energy densification ratio (EDR) in the horse chestnuts, oak acorns, and spruce cones increased as the temperature of the process also increased (Figure 15). This is related to the much higher HHV of the torrefied materials compared to the raw biomass. For the horse chestnuts and oak acorns torrefied at 320 °C, the EDR has the same value of 1.4. The HHV of these materials after torrefaction at 320 °C was 40% higher than the HHV of its raw material. For the spruce cones, the EDR amounted to 1.26. The increase in the EDR testifies to the increase in the HHV. The impact of torrefaction on higher HHV has been determined in many studies. Increases in HHV after torrefaction was observed for food wastes (EDR = 1.57) [10], leaves (EDR = 1.37) [61] and wood (EDR = 1.26) [63].

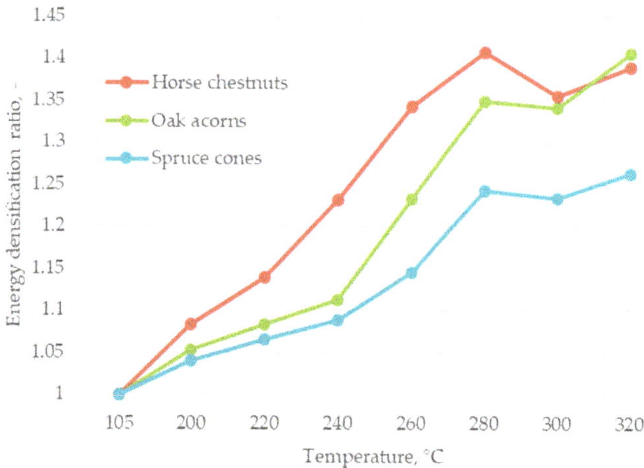

Figure 15. Energy densification ratio in the torrefied forestry biomass.

Based on the mass yield and energy densification ratio, energy yield was determined (Figure 16). The energy yield increased with the process temperature, but reached a maximum at 280 °C. Above this temperature, the EY fell. Forestry biomass torrefied at 320 °C was characterized by the lowest energy yield, even lower than in the raw material, although, the EDR for this temperature was the highest. This can be explained by the fact that energy yield depends on the mass loss, which was the lowest at this temperature. The energy yield factor indicates which temperature of the torrefaction process is the most favorable in terms of energy properties. In analyzing other studies, some differences were observed in energy yield. For torrefied wood and bark, the EY decreased already at the minimal temperature of the process. Energy yield for wood at 280 °C was ca. 11%, and for bark it was 20% [72]. For food wastes (i.e., vegetables, grains, or meats), an initial increase of energy yield was observed, however, at the higher temperature (300 °C) there was a significant decrease in EY. Energy yield between 250 °C and 350 °C decreased from 80% to 40% [73].

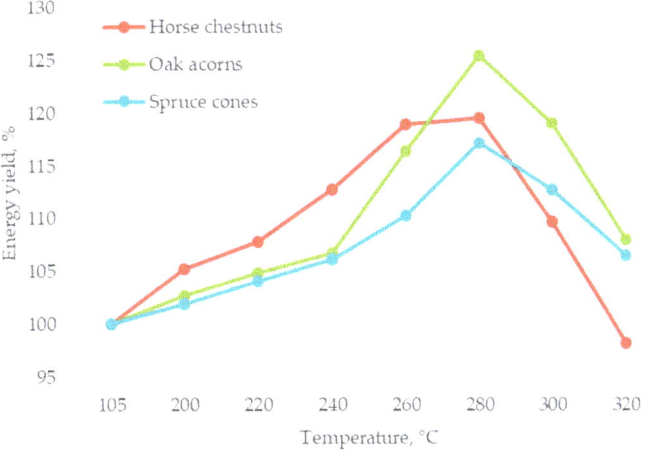

Figure 16. Energy yield of the torrefied forestry biomass.

4. Conclusions

This study proposes thermal processing (torrefaction) of forestry biomass residues (horse chestnuts, oak acorns, and spruce cones) as an initial treatment and a means of preparing alternative fuels or substrates for other applications. The torrefaction process allows for obtaining better fuel properties of raw biomass. Based on the results it can be concluded that horse chestnuts, oak acorns and spruce cones are good organic materials for producing alternative fuel. After the torrefaction process, the tested materials were characterized by very good hydrophobic properties, higher heating value, and higher energy densification. These properties improve the attractiveness of using these materials as fuel. Taking into account many physical and chemical parameters, such as volatile matters content, higher heating value, and fixed carbon content, forestry biomass resembles hard coal after torrefaction, but still remains biomass, which is recognized as an ecological and environmentally friendly source of energy. However, the selection of the torrefaction temperature should be preceded by analysis of the fuel properties. This will help to properly design a valorization process and to save energy inputs as well as financial expenses.

This research provides a starting point for further analysis of horse chestnuts, oak acorns, and spruce cones, which could focus on a more detailed explanation of the different properties and behaviors of these organic matters during thermal processing including, for example, thermogravimetric analysis (TGA), derivative thermogravimetric (DTG) analysis or differential scanning calorimetry (DSC) analysis. Also, from a practical point of view, it is worth examining/determining all the costs involved in the entire logistics chain (harvesting, milling, the torrefaction process, compaction, storage, transport). Furthermore, the utilization of these residues as substrates for other applications (specific chemicals, additives, etc.) would also be interesting from a scientific point of view.

Supplementary Materials: The following are available online at http://www.mdpi.com/1996-1073/13/10/2468/s1, Table S1. Analysis of ash content, volatile matter content, and higher heating value with standard deviation, and coefficient of variation.; Table S2. Analysis of fixed carbon content, and moisture content with standard deviation, and coefficient of variation; Table S3. Analysis of bulk density, specific density, and porosity with standard deviation, and coefficient of variation; Table S4. Analysis of mass yield, energy densification ratio, and energy yield with standard deviation, and coefficient of variation; Table S5. Analysis of hydrophobic properties with standard deviation, and coefficient of variation.

Author Contributions: A.D. conceived and designed the research; T.N. performed the research; A.D. and T.N. analyzed the data and contributed materials/analysis tools; A.D. and T.N. wrote the paper. All authors have read and agreed to the published version of the manuscript.

Funding: This research received no external funding.

Conflicts of Interest: The author declares no conflict of interest.

Abbreviations

AC	ash content
DSC	differential scanning calorimetry
DTG	derivative thermogravimetry
EDR	energy densification ratio
EU	European Union
EY	energy yield
FCC	fixed carbon content
HHV	higher heating value
MC	moisture content
MY	mass yield
RES	renewable energy sources
TGA	thermogravimetric analysis
WDPT	water drop penetration time
VMC	volatile matter content
VOC	volatile organic compound
ρ_B	bulk density
ρ_S	specific density
ε	porosity

References

1. International Energy Agency. *Data and Statistics: Energy Consumption, Electricity Final Consumption*; World. IEA: Paris, France, 2019.
2. Hawkins, E.; Ortega, P.; Suckling, E.; Schurer, A.; Hegerl, G.; Jones, P.; van Oldenborgh, G.J. Estimating Changes in Global Temperature since the Preindustrial Period. *Bull. Am. Meteorol. Soc.* **2017**, *98*, 1841–1856. [CrossRef]
3. Seidl, R.; Thom, D.; Kautz, M.; Martin-Benito, D.; Peltoniemi, M.; Vacchiano, G.; Wild, J.; Ascoli, D.; Petr, M.; Honkaniemi, J.; et al. Forest disturbances under climate change. *Nat. Clim. Chang.* **2017**, *7*, 395–402. [CrossRef] [PubMed]
4. The Intergovernmental Panel on Climate Change. *Special Report: Global Warming of 1.5 °C*; IPCC: Saint-Aubin, France, 2018.
5. Hamelin, L.; Borzęcka, M.; Kozak, M.; Pudełko, R. A spatial approach to bioeconomy: Quantifying the residual biomass potential in the EU-27. *Renew. Sustain. Energy Rev.* **2019**, *100*, 127–142. [CrossRef]
6. Toklu, E. Biomass Energy potential and utilization in Turkey. *Renew. Energy* **2017**, *107*, 235–244. [CrossRef]
7. Dyjakon, A.; Garcia-Galindo, D. Implementing Agricultural Pruning to Energy in Europe: Technical, Economic and Implementation Potentials. *Energies* **2019**, *12*, 1513. [CrossRef]
8. Baldwin, R.F. Forest Products Utilization within a Circular Bioeconomy. *For. Prod. J.* **2020**, *70*, 4–9. [CrossRef]
9. Agbor, V.B.; Cicek, N.; Sparling, R.; Berlin, A.; Levin, D.B. Biomass pretreatment: Fundamentals toward application. *Biotechnol. Adv.* **2011**, *29*, 675–685. [CrossRef]
10. Dyjakon, A.; Noszczyk, T.; Smędzik, M. The Influence of Torrefaction Temperature on Hydrophobic Properties of Waste Biomass from Food Processing. *Energies* **2019**, *12*, 4609. [CrossRef]
11. Tabakaev, R.; Shanenkov, I.; Kazakov, A.; Zavorin, A. Thermal processing of biomass into high-calorific solid composite fuel. *J. Anal. Appl. Pyrol.* **2017**, *124*, 94–102. [CrossRef]

12. Adams, P.; Bridgwater, T.; Lea-Langton, A.; Ross, A.; Watson, I. *Greenhouse Gas Balances of Bioenergy Systems*; Academic Press: Cambridge, MA, USA, 2018; ISBN 978-008-10-1036-5.

13. Qiaoming, L.; Chmely, S.C.; Abdoulmoumine, N. Biomass Treatment Strategies for Thermochemical Conversion. *Energy Fuels* **2017**, *31*, 3525–3536. [CrossRef]

14. Loehle, C. Carbon Sequestration Due to Commercial Forestry: An Equilibrium Analysis. *Forest Prod. J.* **2020**, 60–63. [CrossRef]

15. Szwaja, S.; Poskart, A.; Zajemska, M. A new approach for evaluating biochar quality from Virginia Mallow biomass thermal processing. *J. Clean. Prod.* **2019**, *214*, 356–364. [CrossRef]

16. Chen, D.; Mei, J.; Li, H.; Li, Y.; Lu, M.; Ma, T.; Ma, Z. Combined pretreatment with torrefaction and washing using torrefaction liquid products to yield upgraded biomass and pyrolysis products. *Bioresour. Technol.* **2017**, *228*, 62–68. [CrossRef] [PubMed]

17. Stępień, P.; Pulka, J.; Białowiec, A. Organic Waste Torrefaction–A Review: Reactor Systems, and the Biochar Properties. In *Pyrolysis*; IntechOpen: London, UK, 2017; ISBN 978-953-51-3312-4.

18. Jin, J.; Li, Y.; Zhang, J.; Wu, S.; Cao, Y.; Liang, P.; Zhang, J.; Wong, M.H.; Wand, M.; Shan, S.; et al. Influence of pyrolysis temperature on properties and environmental safety of heavy metals in biochars derived from municipal sewage sludge. *J. Hazard. Mater.* **2016**, *320*, 417–426. [CrossRef] [PubMed]

19. Li, H.; Mahyoub, S.A.A.; Liao, W.; Xia, S.; Zhao, H.; Guo, M.; Ma, P. Effect of pyrolysis temperature on characteristics and aromatic contaminants adsorption behavior of magnetic biochar derived from pyrolysis oil distillation residue. *Bioresour. Technol.* **2017**, *223*, 20–26. [CrossRef]

20. Zhang, Y.; Song, K. Thermal and chemical characteristics of torrefied biomass derived from a generated volatile atmosphere. *Energy* **2018**, *165*, 235–245. [CrossRef]

21. Jagodzińska, K.; Czerep, M.; Kudlek, E.; Wnukowski, M.; Yang, W. Torrefaction of wheat-barley straw: Composition and toxicity of torrefaction condensates. *Biomass Bioenergy* **2019**, *129*, 105335. [CrossRef]

22. Su, Y.; Zhang, S.; Liu, L.; Xu, D.; Xiong, Y. Investigation of representative components of flue gas used as torrefaction pretreatment atmosphere and its effects on fast pyrolysis behaviors. *Bioresour. Technol.* **2018**, *267*, 584–590. [CrossRef]

23. Li, S.X.; Chen, C.Z.; Li, M.F.; Xiao, X. Torrefaction of corncob to produce charcoal under nitrogen and carbon dioxide atmospheres. *Bioresour. Technol.* **2018**, *249*, 348–353. [CrossRef]

24. Chen, W.; Peng, J.; Bi, X. A State of the Art Review of Biomass Torrefaction, Densification and Application. *Renew. Sustain. Energy Rev.* **2015**, *44*, 847–866. [CrossRef]

25. Via, B.K.; Adhikari, S.; Taylor, S. Modeling for proximate analysis and heating value of torrefied biomass with vibration spectroscopy. *Bioresour. Technol.* **2013**, *133*, 1–8. [CrossRef] [PubMed]

26. Quan, C.; Gao, N.; Song, Q. Pyrolysis of biomass components in a TGA and a fixed-bed reactor: Thermochemical behaviors, kinetics, and product characterization. *J. Anal. Appl. Pyrol.* **2016**, *121*, 84–92. [CrossRef]

27. Burhenne, L.; Messmer, J.; Aicher, T.; Laborie, M.P. The effect of the biomass components lignin, cellulose and hemicellulose on TGA and fixed bed pyrolysis. *J. Anal. Appl. Pyrol.* **2013**, *101*, 177–184. [CrossRef]

28. Niu, Y.; Lv, Y.; Lei, Y.; Liu, S.; Liang, Y.; Wang, D.; Hui, S. Biomass torrefaction: Properties, applications, challenges, and economy. *Renew. Sustain. Energy Rev.* **2019**, *115*, 109395. [CrossRef]

29. Cardona, S.; Gallego, L.J.; Valencia, V.; Martinez, E.; Rios, L.A. Torrefaction of eucalyptus-tree residues: A new method for energy and mass balances of the process with the best torrefaction conditions. *Sustain. Energy Techn.* **2019**, *31*, 17–24. [CrossRef]

30. Uslu, A.; Faaij, A.P.C.; Bergman, P.C.A. Pre-Treatment Technologies and their Effect on International Bioenergy Supply Chain Logistics. Techno-Economic Evaluation of Torrefaction, Fast Pyrolysis and Palletisation. *Energy* **2008**, *33*, 1206–1223. [CrossRef]

31. Yang, Y.; Sun, M.; Zhang, M.; Zhang, K.; Wang, D.; Lei, C. A fundamental research on synchronized torrefaction and pelleting of biomass. *Renew. Energy* **2019**, *142*, 668–676. [CrossRef]

32. Acharjee, T.C.; Coronella, C.J.; Vasquez, V.R. Effect of thermal pretreatment on equilibrium moisture content of lignocellulosic biomass. *Bioresour. Technol.* **2011**, *102*, 4849–4854. [CrossRef]

33. Kanwal, S.; Chaudhry, N.; Munir, S.; Sana, H. Effect of torrefaction conditions on the physicochemical characterization of agricultural waste (sugarcane bagasse). *Waste Manag.* **2019**, *88*, 280–290. [CrossRef]

34. Pahla, G.; Ntuli, F.; Muzenda, E. Torrefaction of landfill food waste for possible application in biomass co-firing. *Waste Manag.* **2018**, *71*, 512–520. [CrossRef]

35. Colin, B.; Dirion, J.L.; Arlabosse, P.; Salvador, S. Quantification of the torrefaction effects on the grindability and the hygroscopicity of wood chips. *Fuel* **2017**, *197*, 232–239. [CrossRef]

36. Alvarez, J.G.; Moya, R.; Puente-Urbina, A.; Rodriguez-Zuniga, A. Thermogravimetric, Volatilization Rate, and Differential Scanning Calorimetry Analyses of Biomass of Tropical Plantation Species of Costa Rica Torrefied at Different Temperatures and Times. *Energies* **2018**, *11*, 696. [CrossRef]

37. Singh, R.; Krishna, B.B.; Kumar, J.; Bhaskar, T. Opportunities for Utilization of Non-Conventional Energy Sources for Biomass Pretreatment. *Bioresour. Technol.* **2016**, *199*, 398–407. [CrossRef] [PubMed]

38. PN-EN ISO 18134-2:2017-03E. *Solid Biofuels. Determination of Moisture Content—Oven Dry Method—Part 2: Total Moisture—Simplified Method*; European Committee for Standardization: Brussels, Belgium, 2017.

39. PN-EN ISO 18122:2015. *Solid Biofuels. Determination of Ash Content*; European Committee for Standardization: Brussels, Belgium, 2015.

40. PN-EN ISO 18123:2016-01. *Solid Fuels. Determination of Volatile Content by Gravimetric Method*; European Committee for Standardization: Brussels, Belgium, 2016.

41. PN-EN ISO 18125:2017-07. *Solid Biofuels. Determination of Calorific Value*; European Committee for Standardization: Brussels, Belgium, 2017.

42. ASTM D 3172-73. *Standard Method for Proximate Analysis of Coal and Coke*; ASTM International: Conshohocken, PA, USA, 1984.

43. Doerr, S.H. On Standardizing the "Water Drop Penetration Time" and the "Molarity of An Ethanol Droplet" Techniques to Classify Soil Hydrophobicity: A Case Study Using Medium Textured Soils. *Earth Surf. Process. Landf.* **1998**, *23*, 663–668. [CrossRef]

44. Guatam, R.; Ashwath, N. Hydrophobicity of 43 Potting Media: Its Implications for Raising Seedlings in Revegetation Programs. *J. Hydrol.* **2012**, *430–431*, 111–117. [CrossRef]

45. PN-EN 1237:2000. *Fertilizers—Determination of Bulk Density (Tapped)*; European Committee for Standardization: Brussels, Belgium, 2000.

46. PN-EN 1936:2010. *Natural Stone Test Methods—Determination of Real Density and Apparent Density, and of Total and Open Porosity*; European Committee for Standardization: Brussels, Belgium, 2010.

47. Chin, K.L.; H'ng, P.S.; Go, W.Z.; Wong, W.Z.; Lim, T.W.; Maminski, M.; Paridah, M.T.; Luqman, A.C. Optimization of torrefaction conditions for high energy density solid biofuel from oil palm biomass and fast growing species available in Malaysia. *Ind. Crops Prod.* **2013**, *49*, 768–774. [CrossRef]

48. Chen, W.H.; Lin, B.J.; Colin, B.; Chang, J.S.; Petrissans, A.; Bi, X.; Petrissans, M. Hygroscopic transformation of woody biomass torrefaction for carbon storage. *Appl. Energy* **2018**, *231*, 768–776. [CrossRef]

49. Berther, M.A.; Commandre, J.M.; Rouau, X.; Gontard, N.; Angellier-Coussy, H. Torrefaction Treatment of Lignocellulosic Fibres for Improving Fibre, Matrix Adhesion in a Biocomposite. *Mater. Des.* **2016**, *92*, 223–232. [CrossRef]

50. Wang, Z.; Lim, C.J.; Grace, J.R.; Li, H.; Parise, N.R. Effects of temperature and particle size on biomass torrefaction in a slot-rectangular spouted bed reactor. *Bioresour. Technol.* **2017**, *244*, 281–288. [CrossRef]

51. He, J.; Zhu, L.; Liu, C.; Bai, Q. Optimization of the oil agglomeration for high-ash content coal slime based on design and analysis of response surface methodology (RSM). *Fuel* **2019**, *254*, 115560. [CrossRef]

52. Dyjakon, A.; Noszczyk, T. The influence of freezing temperature storage on the mechanical durability of commercial pellets from biomass. *Energies* **2019**, *12*, 2627. [CrossRef]

53. Jiang, L.; Hu, S.; Xiang, J.; Su, S.; Sun, L.-S.; Xu, K.; Yao, Y. Release characteristics of alkali and alkaline earth metallic species during biomass pyrolysis and steam gasification process. *Bioresour. Technol.* **2012**, *116*, 278–284. [CrossRef]

54. Noda, R.; Matsuhisa, Y.; Ito, T.; Horio, M. Alkali metal evolution characteristics of wood biomass during pyrolysis and gasification. In Proceedings of the Annual Conference of The Japan Institute of Energy, Sapporo, Japan, 30–31 July 2003.

55. Uemura, Y.; Omar, W.N.; Tsutsui, T.; Yusup, S.B. Torrefaction of oil palm wastes. *Fuel* **2011**, *90*, 2585–2591. [CrossRef]

56. Riaza, J.; Gibbins, J.; Chalmers, H. Ignition and combustion of single particles of coal and biomass. *Fuel* **2017**, *202*, 650–655. [CrossRef]

57. Tong, S.; Xiao, L.; Li, X.; Zhu, X.; Liu, H.; Luo, G.; Worasuwannarak, N.; Kerdsuwan, S.; Fungtammasan, B.; Yao, H. A gas-pressurized torrefaction method for biomass wastes. *Energy Convers. Manag.* **2018**, *173*, 29–36. [CrossRef]

58. Correia, R.; Goncalves, M.; Nobre, C.; Mendes, B. Impact of torrefaction and low-temperature carbonization on the properties of biomass wastes from *Arundo donax* L. and *Phoenix canariensis*. *Bioresour. Technol.* **2017**, *223*, 210–218. [CrossRef]

59. Bai, Z.; Liu, Q.; Lei, J.; Hong, H.; Jin, H. New-solar biomass power generation system integrated a two-stage gasifier. *Appl. Energy* **2017**, *194*, 310–319. [CrossRef]

60. Geng, C.; Yang, W.; Sun, X.; Wang, X.; Bai, Z.; Zhang, X. Emission factors, ozone and secondary organic aerosol formation potential of volatile organic compounds emitted from industrial biomass boilers. *Int. J. Environ. Sci.* **2019**, *83*, 64–72. [CrossRef]

61. Conag, A.T.; Villahermosa, J.E.R.; Cabatingan, L.K.; Go, A.W. Energy densification of sugarcane leaves through torrefaction under minimized oxidative atmosphere. *Energy Sustain. Dev.* **2018**, *42*, 160–169. [CrossRef]

62. Uzun, H.; Yildiz, Z.; Goldfarb, J.L.; Ceylan, S. Improved prediction of higher heating value of biomass using an artificial neural network model based on proximate analysis. *Bioresour. Technol.* **2017**, *234*, 122–130. [CrossRef]

63. Świechowski, K.; Liszewski, M.; Bąbelewski, M.; Koziel, J.A.; Białowiec, A. Fuel properties of torrefied biomass from pruning of oxytree. *Data* **2019**, *4*, 55. [CrossRef]

64. Tumuluru, J.S.; Sokhansanj, S.; Wright, C.T.; Boardman, R.D.; Hess, R.J. Review on Biomass Torrefaction Process and Product Properties and Design of Moving Bed Torrefaction System Model Development. In Proceedings of the ASABE Annual International Meeting, Louisville, KY, USA, 7–10 August 2011.

65. Chen, Y.; Liu, B.; Yang, H.; Yang, Q.; Chen, H. Evolution of Functional Groups and Pore Structure During Cotton and Corn Stalks Torrefaction and its Correlation with Hydrophobicity. *Fuel* **2014**, *137*, 41–49. [CrossRef]

66. Alvarez, A.; Gutierrez, G.; Matos, M.; Pizarro, C.; Bueno, J.L. Torrefaction of short rotation coppice of poplar under oxidative and non-oxidative atmosphere. *Multidiscip. Digit. Publ. Inst. Proc.* **2018**, *2*, 1479. [CrossRef]

67. Pouzet, M.; Dubois, M.; Charlet, K.; Petit, E.; Beakou, A.; Dupont, C. Fluorination/Torrefaction Combination to Further Improve the Hydrophobicity of Wood. *Macromol. Chem. Phys.* **2019**, *220*. [CrossRef]

68. Piccand, M.; Bianchi, S.; Halaburt, E.I.; Mayer, I. Characterization of extractives from biomasses of the alpine forests and their antioxidative efficacy. *Ind. Crops. Prod.* **2019**, *142*, 111832. [CrossRef]

69. Eberhardt, T.L.; Han, J.S.; Micales, J.A.; Young, R.A. Decay Resistance in Conifer Seed Cones: Role of Resin Acids as Inhibitors of Decomposition by White-Rot Fungi. *Holzforschung* **1994**, *48*, 278–284. [CrossRef]

70. Nhuchhen, D.R.; Basu, P.; Acharya, B. A comprehensive Review on Biomass Torrefaction. *IJREB* **2014**. [CrossRef]

71. Bach, Q.V.; Skreiberg, O. Upgrading biomass fuels via wet torrefaction: A review and comparison with dry torrefaction. *Renew. Sustain. Energy Rev.* **2016**, *54*, 665–677. [CrossRef]

72. Almeida, G.; Brito, J.O.; Perre, P. Alterations in energy properties of eucalyptus wood and bark subjected to torrefaction: The potential of mass loss as a synthetic indicator. *Bioresour. Technol.* **2010**, *101*, 9778–9784. [CrossRef]

73. Poudel, J.; Ohm, T.I.; Oh, S.C. A study on torrefaction of food waste. *Fuel* **2015**, *140*, 275–281. [CrossRef]

Article

Waste to Carbon: Biocoal from Elephant Dung as New Cooking Fuel

Paweł Stępień [1], Kacper Świechowski [1], Martyna Hnat [1], Szymon Kugler [2], Sylwia Stegenta-Dąbrowska [1,*], Jacek A. Koziel [3], Piotr Manczarski [4] and Andrzej Białowiec [1,3]

[1] Institute of Agricultural Engineering, Faculty of Life Sciences and Technology, Wroclaw University of Environmental and Life Sciences, 37a Chełmońskiego Str., 51-630 Wroclaw, Poland; pawel.stepien@upwr.edu.pl (P.S.); kacper.swiechowski@upwr.edu.pl (K.Ś.); hnat.martyna@gmail.com (M.H.); andrzej.bialowiec@upwr.edu.pl (A.B.)

[2] Polymer Institute, Faculty of Chemical Technology and Engineering, West Pomeranian University of Technology, 10 Pułaskiego Str., 70-322 Szczecin, Poland; szymon.kugler@zut.edu.pl

[3] Department of Agricultural and Biosystems Engineering, Iowa State University, Ames, IA 50011-3270, USA; koziel@iastate.edu

[4] Department of Environmental Engineering, Hydro and Environmental Engineering, Faculty of Building Services, Warsaw University of Technology, 00-661 Warszawa, Poland; piotr.manczarski@pw.edu.pl

* Correspondence: sylwia.stegenta@upwr.edu.pl; Tel.: +48-71-320-5811

Received: 10 September 2019; Accepted: 12 November 2019; Published: 14 November 2019

Abstract: The paper presents, for the first time, the results of fuel characteristics of biochars from torrefaction (a.k.a., roasting or low-temperature pyrolysis) of elephant dung (manure). Elephant dung could be processed and valorized by torrefaction to produce fuel with improved qualities for cooking. The work aimed to examine the possibility of using torrefaction to (1) valorize elephant waste and to (2) determine the impact of technological parameters (temperature and duration of the torrefaction process) on the waste conversion rate and fuel properties of resulting biochar (biocoal). In addition, the influence of temperature on the kinetics of the torrefaction and its energy consumption was examined. The lab-scale experiment was based on the production of biocoals at six temperatures (200–300 °C; 20 °C interval) and three process durations of the torrefaction (20, 40, 60 min). The generated biocoals were characterized in terms of moisture content, organic matter, ash, and higher heating values. In addition, thermogravimetric and differential scanning calorimetry analyses were also used for process kinetics assessment. The results show that torrefaction is a feasible method for elephant dung valorization and it could be used as fuel. The process temperature ranging from 200 to 260 °C did not affect the key fuel properties (high heating value, HHV, HHV_{daf}, regardless of the process duration), i.e., important practical information for proposed low-tech applications. However, the higher heating values of the biocoal decreased above 260 °C. Further research is needed regarding the torrefaction of elephant dung focused on scaling up, techno-economic analyses, and the possibility of improving access to reliable energy sources in rural areas.

Keywords: torrefaction; biorenewable energy; biowaste; biocoal; alternative fuel; waste management; manure; thermal valorization; thermogravimetric analysis; differential scanning calorimetry

---

1. Introduction

It is estimated that there are around~450,000 elephants today, of which 400,000 are in Africa and 50,000 in Asia. In Africa, these mammals live in 34 countries (Angola, Benin, Botswana, Burkina Faso, Cameroon, Central African Republic, Chad, Congo, Ivory Coast, Equatorial Guinea, Eritrea, Ethiopia, Gabon, Gana, Guinea, Bissau, Kenya, Liberia, Malawi, Mali, Mozambique, Namibia, Niger, Nigeria, Rwanda, Senegal, Sierra Leone, Somalia, South Africa, Sudan, Tanzania, Togo, Uganda, Zambia,

Zimbabwe), and on the Asian continent they can be found in 15 countries (India, Nepal, Bhutan and Bangladesh, China, Burma, Thailand, Cambodia, Laos, Vietnam, Malaysia, Andaman Islands, Sri Lanka, Sumatra, Borneo) [1]. The daily amount of dung produced by one elephant is 100–150 kg. The weight of elephant excrement depends on the amount of consumed water [2–4]. Thus, taking into consideration the conservative estimate of the minimum dung weight (100 kg), the daily and annual dung production on a global scale is 45,000 Mg and more than 16 million Mg, respectively, i.e., a large amount of biowaste that could be valorized [2–4].

From an ecological point of view, untreated animal waste or handling, air-drying and combustion without prior treatment can be problematic due to health and environmental concerns, such as elevated risk of contamination with pathogens, contamination of drinking water sources, gaseous emissions of odor, hydrogen sulfide, ammonia, and other toxic gases [5,6]. In addition, the loss of nutrients from dung associated with current practices can also represent economic losses due to its lower value as a fertilizer [5].

We propose a solution to these problems with the introduction of the torrefaction process to manage and valorize the elephant dung. Resulting biocoal can be used as a fuel with a useful high heating value (*HHV*). Research with slow pyrolysis and hydrothermal carbonization of other types of livestock manure resulted in HHVs ranging from 15.8 to 18.4 MJ/kg [7]. Qambrani et al. [8] showed that biocoal from animal manure contains more N compared to biochar from plant residues. Although the pore structure is more organized in biochar from plant sources, the fertilizer quality and heavy metal adsorbability were found to be excellent in manure biochars. On the other hand, some raw waste types (such as poultry manure or sewage sludge) can contain a large amount of copper and zinc, which limits its use as a fertilizer. The proposed concept to valorize elephant manure can provide new technologies for using the torrefaction process in rural areas, which can be used to obtain better quality fuel and fertilizer.

To date, several methods to valorize elephant dung have been proposed. Vermicomposting is a biological process in which the organic fraction of dung is decomposed by microorganisms and earthworms under controlled environmental conditions to a level when it can be applied on arable land. This method can be ecological and economically profitable [9]. Vermicomposting of animal dung from the zoo was investigated in pilot-scale by a team of scientists in Mexico [6]. Elephant dung was also used for research by scientists in Thailand for the production of biogas in co-fermentation with water hyacinth and fermentation on a laboratory scale. In the case of co-fermentation, the calorific value of biogas was 15.05 MJ·m^{-3} [10,11].

Biohydrogen production through anaerobic mixed cultures of microorganisms found in elephant dungs was also researched in laboratory conditions. It is based on simultaneous saccharification and fermentation of cellulose. The bacteria break down the cellulose to glucose, and then non-cellulolytic bacteria from the formed glucose produces hydrogen [12,13]. The microorganism's culture from elephant dung stimulated the production of H$_2$ from cellulose. It was assumed that cellulolytic bacteria in the dung originate from the plant diet of the elephant. Animal manure, including elephant dung, was also the subject of research conducted in Thailand on cellulolytic bacteria for the direct production of butanol from cellulose, which could be an alternative to fuel obtained from petroleum [14].

The knowledge about practical considerations for the valorization of elephant dung and the progression from lab to full-scale (e.g., costs of construction and operation) is limited. There are also questions about the storage and distribution of finished products (e.g., fuel briquettes for cooking), which could be prohibitively expensive for long-range transport. Life-cycle analyses could be useful to assess the critical transport range [15]. It is equally important to consider managing the residues (e.g., raw dung and sludge), which may require specialized collection, storage, treatment, and disposal. It has not been described yet how existing or developing technologies (anaerobic digestion, biohydrogen production) could be used for waste management, especially in rural regions in which elephant dung is available in large quantities. Thus, there is a need to find local-scale solutions suited for these regions, which should be safe, inexpensive, simple to build, use and maintain, dependable, and not generating another waste stream to manage.

We propose an alternative solution for elephant dung management via torrefaction (Figure 1). Torrefaction (a.k.a., 'roasting' or low-temperature pyrolysis) is a thermochemical process occurring at 200–300 °C without the presence of an oxidant. Jia et al. [16] described the possibility of using a co-gasification of woody biomass and animal manure as a useful technology to utilize organic waste, which could be practical in the case of elephant dung as well. The elephant dung fuel produced may be an attractive source of rural fuel. For example, in India alone, 6.3% of all households use the so-called 'dung cake' to produce the energy needed for cooking [17]. Assuming 1.34 billion people in India in 2018 [18] and that one household comprises 10 people, as many as ~8.4 million households use dung cake for energy production. Although the torrefaction process requires some energy, it is also the most promising technology for organic waste treatment for its highest greenhouse gas mitigation potential [19]. The produced biocoal, especially when pelletized, poses a lower environmental risk during transport, storage, and combustion, in addition to lowering the risks of sanitary and aquatic pollution [20,21]. Therefore, torrefaction could be one of the potential technologies for elephant dung utilization that are sustainable.

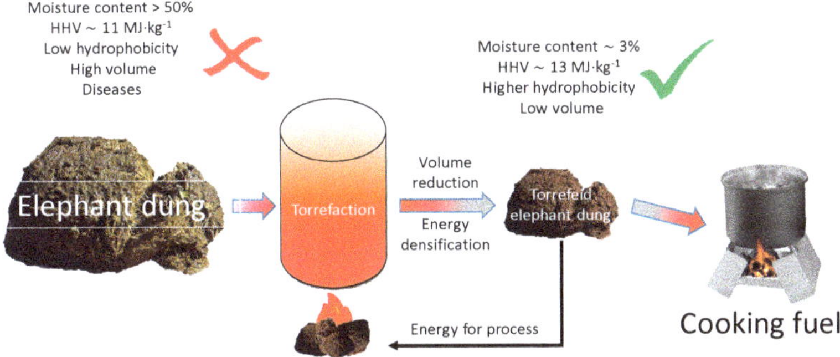

Figure 1. The proposed valorization of elephant dung (manure) via torrefaction.

To date, no work has been carried out on the torrefaction of elephant dung as a method for the production of fuel. Local-scale torrefaction can address challenges with dung management, through its valorization, while improving the socio-economic situation in rural households. Therefore, the research carried out was aimed at determining:

- Whether torrefaction can be used as a method of preliminary valorization of elephant dung;
- Whether the duration of the torrefaction process at a given temperature affects the dung conversion rate (e.g., mass loss, energy densification, and improved fuel properties);
- Whether energy consumption is needed for the torrefaction of elephant dung.

2. Materials and Methods

2.1. Feedstock

The study used Asian elephant dung from the Zoological Garden, located in Wrocław, Poland. The 5 kg sample was dried at 105 °C for 24 h in a laboratory dryer, followed by milling to the grain size of ≤0.425 mm with the laboratory knife mill (TESTCHEM, model LMN-100, Pszów, Poland) to make the sample homogeneous. Samples were frozen at −15 °C for further testing.

2.2. Biocoal Production Method via Torrefaction

A scheme of the experiment is shown in Figure 2. The biocoal production process was carried out in triplicates according to the methodology presented by [22] at six temperatures from 200 to 300 °C

(20 °C intervals) at 20, 40, 60 min for each interval, followed by the cooling phase. The biocoals were generated using a muffle furnace (Snol, model 8.1/1100, Utena, Lithuania). CO_2 inert gas was provided to the furnace to ensure non-oxidative conditions. The elephant dung samples were heated from 20 °C to set point at 50 °C·min⁻¹. The cooling times were 38 min, 33 min, 29 min, 23 and 13.5 min, from torrefaction setpoints of 300 °C, 280 °C, 260 °C, 240 °C, and 220 °C to 200 °C, respectively. After the CO_2 supply was cut off, the biocoals were removed from the furnace when the interior temperature was <200 °C. The mass of the sample was determined before and after the cooling process in order to calculate the mass loss. Dung samples of 10 ± 0.5 g (dry mass, d.m.) were used to produce biocoal.

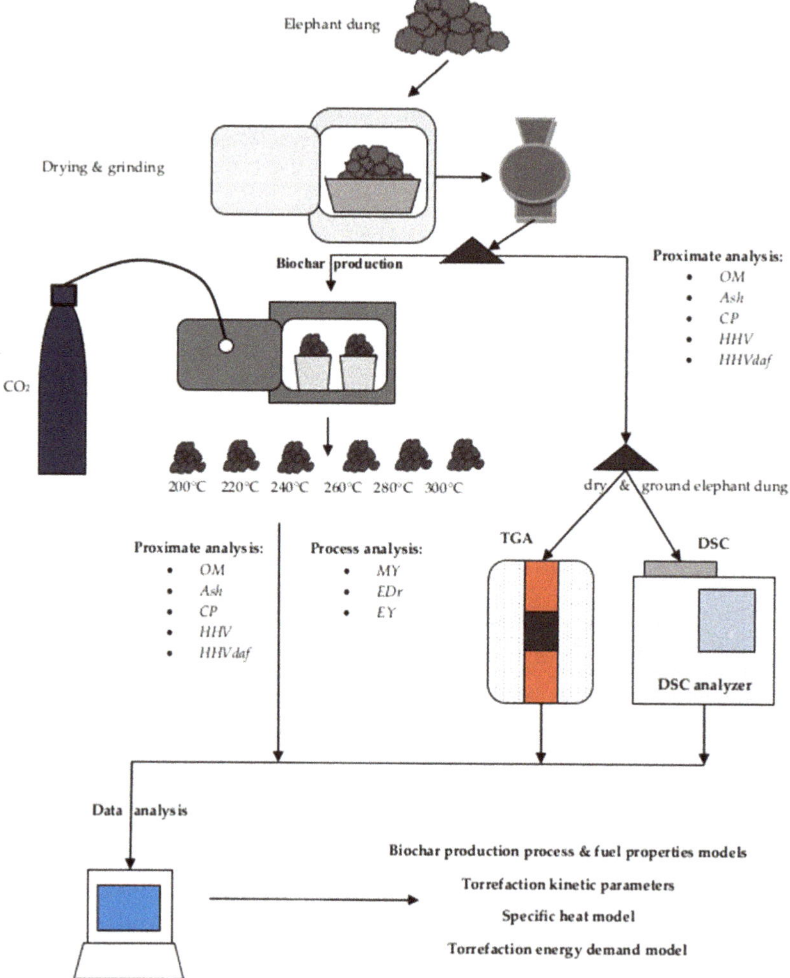

Figure 2. Scheme of experiments – biocoal production via torrefaction of elephant dung to determine the process kinetics with thermogravimetric analyses (TGA) and differential scanning calorimetry (DSC).

2.3. Proximate Analysis of Raw and Torrefied Elephant Dung

Physical and chemical properties were subjected to raw material and produced biocoals. The following tests were made in three replicates using the following standard methods:

- Moisture content (*MC*) by means of a laboratory dryer (WAMED, model KBC-65W, Warsaw, Poland) at temperature 105 °C, time 24 h, in accordance with the PN-EN 14346:2011 standard [23],
- Organic matter content (*OM*) by means of a laboratory dryer (WAMED, model KBC-65W, Warsaw, Poland) at temperature 550 °C, time 4 h, in accordance with the PN-EN 15169:2011 standard [24],
- ash and combustible parts (*CP*) by means of a laboratory dryer (WAMED, model KBC-65W, Warsaw, Poland) at temperature 815 °C, time 4 h in accordance with the PN-Z-15008-04:1993 standard [25],
- High heating value (*HHV*) by means of the IKA C2000 Basic calorimeter, at 17–25 °C, 30 bar pressure in accordance with the PN-G-04513:1981 standard [26].

2.4. Thermogravimetric Analysis (TGA) of Elephant Dung

Thermogravimetric analyses (TGA) were first performed in isothermal conditions to determine the kinetics parameters (*k*—reaction rate constants and E_a—activation energy) of the torrefaction process of elephant dung. Reaction rate constants were determined for the following temperatures: 200 °C, 220 °C, 240 °C, 260 °C, 280 °C, and 300 °C in accordance with the methodology and reactor set-up presented elsewhere [22]. First, the empty furnace was pre-heated to the set point. Then, 3 g of dry dung was placed in the steel crucible and placed in the furnace for 1 h. Measurement of mass loss was performed using a balance coupled to a steel crucible at 10 s intervals with 0.01 g accuracy. The calculating methodology for the kinetic parameters is presented in Section 2.6.2. The kinetics parameters (reaction rate and activation energy) were calculated.

TGA analyses were also completed in non-isothermal conditions to obtain more comprehensive data on the thermal degradation of elephant dung. These TGA analyses were performed at rising temperatures (from 20 °C to 850 °C) at a heating rate of 650 °C·h^{-1} (10.83 °C·min^{-1}). The sample was heated for 2 min after reaching a set point. The study of kinetic parameters and thermal degradation was performed using the stand-mounted tubular furnace (Czylok, RST 40x200/100, Jastrzębie-Zdrój, Poland).

2.5. Differential Scanning Calorimetry (DSC) of Raw Elephant Dung

Differential scanning calorimetry (DSC) analysis was carried out using a differential scanning calorimeter (TA Instruments, DSC Q2500, New Castle, DE, USA). Approximately 6 mg of the tested material was weighed into the aluminum hermetic crucible. Each sample (n = 1) was then placed in the analyzer and heated from 10 °C to 300 °C at a heating rate of 10 °C·min^{-1}. The N_2 inert gas was supplied at 3 dm^3·h^{-1} flowrates. The analysis provided information on endothermic and exothermic changes during torrefaction.

2.6. Data-Processing Calculation Methods

2.6.1. Mass Yield, Energy Densification Ratio, and Energy Yield

The mass yield, energy densification ratio, and energy yield of each of the variants were determined based on Equations (1)–(3), respectively [27]:

$$MY = \frac{m_b}{m_a} \cdot 100 \tag{1}$$

where:

MY—mass yield, %
m_a—the mass of dry elephant dung before torrefaction, g,
m_b—the mass of dry biocoal after torrefaction, g.

$$EDr = \frac{HHV_b}{HHV_a} \tag{2}$$

where:

EDr—energy densification ratio, -,
HHV_b—the high heating value of biocoal, J·g^{-1},
HHV_a—the high heating value of raw elephant dung, MJ·kg^{-1}.

$$EY = MY \cdot ED_r \tag{3}$$

where:

EY—nergy yield, %,
MY—mass yield, %
ED_r—energy densification ratio, -,

The ash-free value of the HHV was determined based on [28]:

$$HHV_{daf} = \frac{HHV}{M_f - M_{ash}} \tag{4}$$

where:

HHV_{daf}—high heating value on dry and ash-free base, MJ·kg^{-1},
HHV—high heating value, MJ·kg^{-1},
M_f—dry mass of fuel, kg,
M_{ash}—the mass of ash in fuel, kg.

2.6.2. Calculation of Kinetics Parameters (Reaction Rate and Activation Energy)

The data obtained from isothermal TGA analysis were used to determine the reaction rate (k) constant for each temperature, based on the first-order model [22]:

$$m_s = m_0 \cdot e^{-k \cdot t} \tag{5}$$

where:

m_s—mass after time t, g,
m_0—initial mass, g,
k—the reaction rate constant, s^{-1},
t—time, s.

The nonlinear estimation of k in Equation (5) for each temperature was made with the Statistica 13.3 software (StatSoft, Inc., TIBCO Software Inc. Palo Alto, CA, USA). The Arrhenius plot was created (ln(k)(T) vs. 1/T) on the basis of k values for individual temperatures [29], and a trend line was found:

$$y = a \cdot x + b \tag{6}$$

Then, the activation energy (E_a) values [22] were determined as follows:

$$E_a = -(a \cdot R) \tag{7}$$

where:

E_a—activation energy, J·mol^{-1},
a—the coefficient from Equation (6), K,
R—gas constant, J·mol^{-1}·K^{-1}.

2.6.3. Calculation of Energy Demand for Torrefaction of Elephant Dung

The results from the DSC and TGA analyses were used to calculate the actual energy demand in processing dry elephant dung (to heat dung from 20 °C to 300 °C) in accordance with the methodology presented in a previous paper [30]. The lack of TGA analysis causes overestimated energy amount needed to process the material, due to the decreasing amount of material during torrefaction caused by its devolatilization. The following is an example of the model use where the calculation for 1 g of the raw elephant dung torrefied at 300 °C was considered. The total amount of energy needed to processing raw elephant dung was calculated by adding the energy needed to evaporate water from raw elephant dung to the result from the model of dry elephant dung. The energy needed to evaporate water was calculated by Equation (8) [31]:

$$Q = m \cdot \Delta T \cdot cp + m \cdot co \tag{8}$$

where:

Q—the total amount of heat needed to heat and evaporate water, J,

m—the mass of water in the sample, g,

ΔT—the temperature difference between ambient temperature (20 °C) and boiling point (100 °C), under normal pressure conditions, °C,

cp—specific heat of water, 4.2 J·(g·°C)$^{-1}$,

co—the heat of water evaporation, 2257 J·g^{-1}.

2.6.4. Modeling of Torrefaction Process and Biocoal Fuel Properties

Polynomial models of the influence of torrefaction temperature and time on torrefaction process and biocoals fuel parameters were developed. These models were based on measured data from the torrefaction process, and biocoal properties for a particular temperature and time using a similar modeling approach described in our previous work [32]. Equations describing *MY, EDr, EY*, organic matter content, combustible parts, ash, *HHV*, and *HHV*$_{daf}$ for biocoal were developed. The general form of the applied polynomial equation was:

$$f(T,t) = a_1 + a_2 \cdot T + a_3 \cdot T^2 + a_4 \cdot t + a_5 \cdot t^2 + a_6 \cdot T \cdot t + a_7 \cdot T^2 \cdot t^2 \tag{9}$$

where:

f(T,t)—the property (*T, t*, & combinations) being analyzed,

a_1—intercept,

a_2–a_7—regression coefficient,

T—process temperature, °C,

t—process time, min.

Regression analysis used a 2-degree polynomial with a general form, with intercept (a_1) and six regression coefficients (a_2–a_7). The confidence interval of the parameter evaluations (a_1–a_7) was 95%. All parameters for which the results of *p*-value were <0.05, were assumed to be statistically significant. The results of the analysis are presented in the form of equations. as well as the correlation coefficients (R) and determination coefficients (R^2). The results of the DSC analysis were also subjected to polynomial regression analysis in order to determine a useful model of the specific heat (SH) of elephant dung for 200–300 °C. The polynomial regression analysis was used because the torrefaction process has a non-linear character. The results were presented in the form of an equation describing the dependence of the change of specific heat of elephant dung as a function of temperature. The general form of the polynomial used is in the form of Equation (10). Nine regression coefficients were used to provide a higher level of matching model to raw data.

$$SH = a_1 + a_2 \cdot T + a_3 \cdot T^2 + a_4 \cdot T^3 + a_5 \cdot T^4 + a_6 \cdot T^5 + a_7 \cdot T^6 + a_8 \cdot T^7 + a_9 \cdot T^8 \tag{10}$$

where:

SH—specific heat of elephant dung as a function of temperature, $J \cdot (kg \cdot °C)^{-1}$,
a_1—intercept,
a_2–a_9—regression coefficient,
T—torrefaction temperature, °C.

Nonlinear regression and evaluation of intercepts and regressions coefficients ($p < 0.05$) were completed with Statistica software (13.3, StatSoft, Palo Alto, CA, USA).

2.6.5. Statistical Analysis

An analysis of variance (ANOVA) evaluation of differences between mean values was performed with the application of post-hoc Tuckey's test, at the $p < 0.05$ significance level. For statistical data evaluation, the Statistica software (13.3, StatSoft, Palo Alto, CA, USA) was used.

3. Results

3.1. Result of the Torrefaction Process

The mass yields (MY) for elephant dung biocoals (Figure 3) showed a downward trend with the increase of process temperature. The highest mass yields values were obtained for biocoal generated at 200 °C and were above 90%. The lowest *MY* was for 300 °C, in this case, the mass yield decreased to 66%. All regression coefficients were statistically significant ($p < 0.05$) in the *MY* model, ($R^2 = 0.75$) (Table 1). Detailed *MY* data are shown in Table A2.

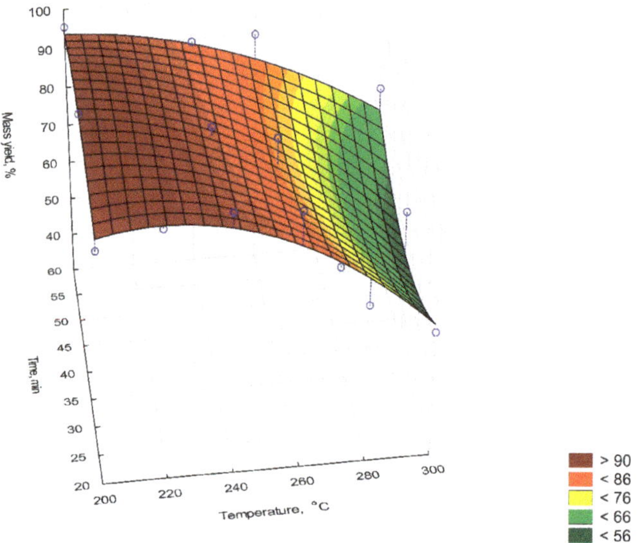

Figure 3. The influence of temperature and time on the mass yield of biocoal from elephant dung.

Table 1. Statistical evaluation of mass yield of biocoal from elephant dung.

Intercept/ Coefficient	Value of Intercept/ Coefficient	Standard Error	p	Lower Limit of Confidence	Upper Limit of Confidence
a_1	-2.58×10^2	2.18×10^2	0.00	-7.38×10^2	2.22×10^2
a_2	2.72×10^0	1.44×10^0	0.00	-4.52×10^{-1}	5.89×10^0
a_3	-5.12×10^{-3}	2.39×10^{-3}	0.00	-1.04×10^{-2}	1.41×10^{-4}
a_4	6.91×10^0	6.19×10^0	0.00	-6.71×10^0	2.05×10^1
a_5	-4.22×10^{-2}	4.01×10^{-2}	0.00	-1.30×10^{-1}	4.61×10^{-2}
a_6	-3.01×10^{-2}	2.45×10^{-2}	0.00	-8.40×10^{-2}	2.38×10^{-2}
a_7	7.70×10^{-7}	0.00×10^0	0.00	7.70×10^{-7}	7.70×10^{-7}

$MY = a_1 + a_2 \cdot T + a_3 \cdot T^2 + a_4 \cdot t + a_5 \cdot t^2 + a_6 \cdot t + a_7 \cdot T^2 \cdot t$, $R^2 = 0.75$, $R = 0.87$; T^* ranged from 200 °C to 300 °C, t^* ranged from 20 min to 60 min; * more information in Section 2.2.

The energy yield (EY) of the biochar from elephant dung (Figure 4) also decreased with the increase of temperature and did not change with time. The biocoals produced at 200 °C resulted in more than 105% EY compared to raw material. However, the EY dropped below 68% for torrefaction at 300 °C. All regression coefficients were statistically significant ($p < 0.05$) for the EY model ($R^2 = 0.85$) (Table 2).

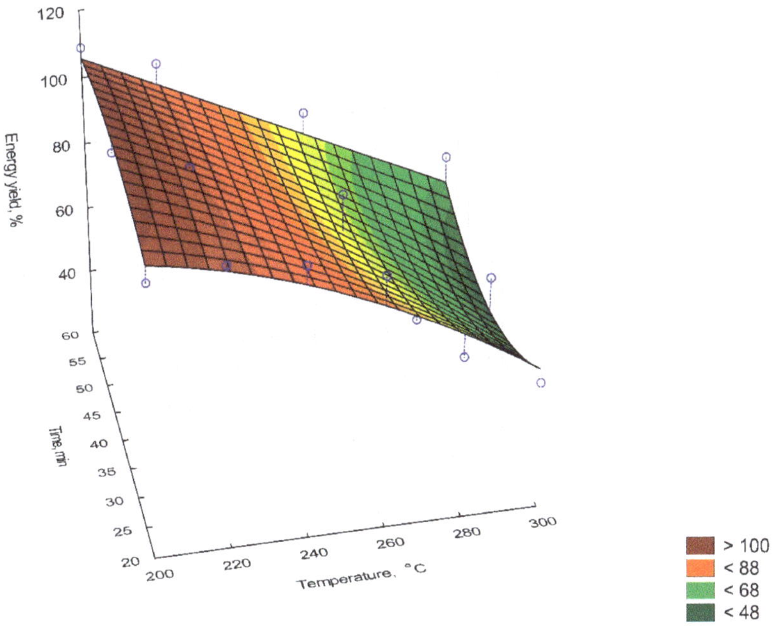

Figure 4. The influence of temperature and time on the energy yield in biocoal from elephant dung.

The energy densification ratio (EDr) in biocoals generated from elephant dung (Figure 5) decreased with increasing temperature and did not change much with time. Biocoals produced at 200 °C had the highest EDr of ~1.1, while biocoals generated at 300 °C had the lowest EDr (~0.9). All regression coefficients were statistically significant ($p < 0.05$) for the EDr model ($R^2 = 0.83$) (Table 3).

Table 2. Statistical evaluation of energy yield of biocoal from elephant dung.

Intercept/ Coefficient	Value of Intercept/ Coefficient	Standard Error	p	Lower Limit of Confidence	Upper Limit of Confidence
a_1	-1.19×10^2	2.48×10^2	0.00	-6.65×10^2	4.27×10^2
a_2	1.69×10^0	1.64×10^0	0.00	-1.91×10^0	5.30×10^0
a_3	-3.17×10^{-3}	2.72×10^{-3}	0.00	-9.16×10^{-3}	2.82×10^{-3}
a_4	8.26×10^0	7.04×10^0	0.00	-7.23×10^0	2.38×10^1
a_5	-4.74×10^{-2}	4.56×10^{-2}	0.00	-1.48×10^{-1}	5.30×10^{-2}
a_6	-3.65×10^{-2}	2.79×10^{-2}	0.00	-9.78×10^{-2}	2.48×10^{-2}
a_7	8.73×10^{-7}	0.00×10^0	0.00	8.73×10^{-7}	8.73×10^{-7}

$EY = a_1 + a_2{\cdot}T + a_3{\cdot}T^2 + a_4{\cdot}t + a_5{\cdot}t^2 + a_6{\cdot}T{\cdot}t + a_7{\cdot}T^2{\cdot}t$, $R^2 = 0.85$, $R = 0.92$; T^* ranged from 200 °C to 300 °C, t^* ranged from 20 min to 60 min; * more information in Section 2.2.

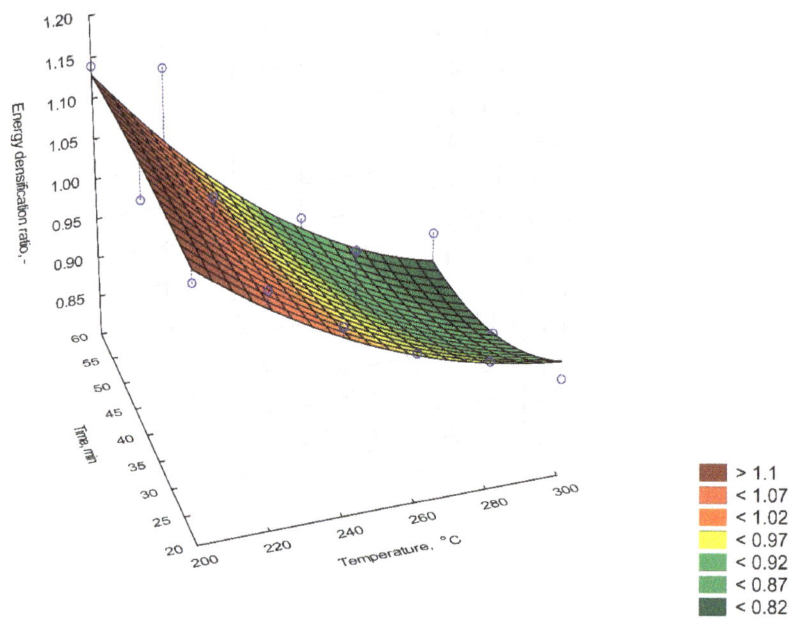

Figure 5. The influence of temperature and time on the energy densification ratio in biocoal from elephant dung.

Table 3. Statistical evaluation of energy densification ratio of biocoal from elephant dung.

Intercept/ Coefficient	Value of Intercept/ Coefficient	Standard Error	p	Lower Limit of Confidence	Upper Limit of Confidence
a_1	1.99×10^0	1.38×10^0	0.00	-1.05×10^0	5.04×10^0
a_2	-7.21×10^{-3}	9.14×10^{-3}	0.00	-2.73×10^{-2}	1.29×10^{-2}
a_3	1.41×10^{-5}	1.52×10^{-5}	0.00	-1.93×10^{-5}	4.75×10^{-5}
a_4	2.23×10^{-2}	3.93×10^{-2}	0.00	-6.41×10^{-2}	1.09×10^{-1}
a_5	-9.11×10^{-5}	2.54×10^{-4}	0.00	-6.51×10^{-4}	4.69×10^{-4}
a_6	-1.06×10^{-4}	1.55×10^4	0.00	-4.48×10^{-4}	2.36×10^{-4}
a_7	1.90×10^{-9}	0.00×10^0	0.00	1.90×10^{-9}	1.90×10^{-9}

$EDr = a_1 + a_2{\cdot}T + a_3{\cdot}T^2 + a{\cdot}t + a_5{\cdot}t^2 + a_6{\cdot}t + a_7{\cdot}T^2 t$, $R^2 = 0.83$, $R = 0.91$; T^* ranged from 200 °C to 300 °C, t^* ranged from 20 min to 60 min; * more information in Section 2.2.

3.2. Result of Proximate Analysis of Raw and Torrefied Elephant Dung

The content of organic matter (*OM*) decreased as the temperature and the retention time increased. The lowest *OM* value was 28.26% for torrefaction at 280 °C and 60 min, and for torrefaction at 300 °C in time 20 min and 40 min (Figure 6, Table A1). Analysis of variance showed that statistically significant differences occur between the results obtained at 260 °C, 280 °C, and 300 °C, ($p < 0.05$) (Figure A1, Table A3). All regression coefficients were statistically significant ($p < 0.05$) for the *OM* model ($R^2 = 0.83$) (Table 4).

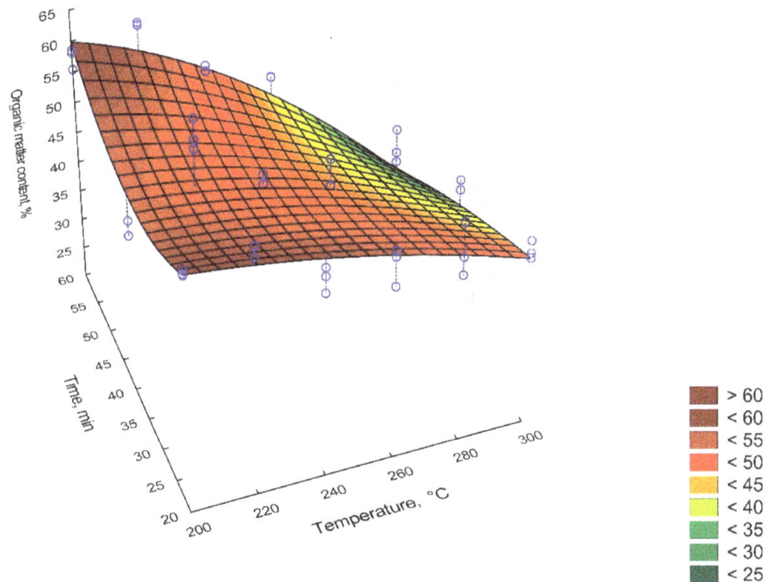

Figure 6. The influence of temperature and time on the organic matter content in biocoal from elephant dung.

Table 4. Statistical evaluation of organic matter content of biocoal from elephant dung.

Intercept/ Coefficient	Value of Intercept/ Coefficient	Standard Error	p	Lower Limit of Confidence	Upper Limit of Confidence
a_1	9.74×10^1	6.33×10^1	0.00	-3.00×10^1	2.25×10^2
a_2	1.24×10^{-2}	4.18×10^{-1}	0.00	-8.29×10^{-1}	8.54×10^{-1}
a_3	-4.95×10^{-4}	6.95×10^{-4}	0.00	-1.89×10^{-3}	9.03×10^{-4}
a_4	-3.25×10^0	1.80×10^0	0.00	-6.87×10^0	3.62×10^{-1}
a_5	3.20×10^{-2}	1.16×10^{-2}	0.00	8.53×10^3	5.54×10^{-2}
a_6	9.62×10^{-3}	7.11×10^{-3}	0.00	-4.69×10^{-3}	2.39×10^{-2}
a_7	-3.85×10^{-7}	0.00×10^0	0.00	-3.85×10^{-7}	-3.85×10^7

$OM = a_1 + a_2 \cdot T + a_3 \cdot T^2 + a_4 \cdot t + a_5 \cdot t^2 + a_6 \cdot T \cdot t + a_7 \cdot T^2 \cdot t$, $R^2 = 0.83$, $R = 0.91$; T^* ranged from 200 °C to 300 °C, t^* ranged from 20 min to 60 min; * more information in Section 2.2.

The ash content was inversely proportional to the *OM* content and increased to over 71% in comparison to 50.81% for raw dung (Table A1) in biocoal produced at 280 and 300 °C at 60 min (Figure 7). Analysis of variance showed statistically significant differences between the results for temperatures 260 °C, 280 °C, and 300 °C ($p < 0.05$), (Figure A2, Table A4). All regression coefficients were statistically significant ($p < 0.05$) for the ash content model ($R^2 = 0.83$) (Table 5).

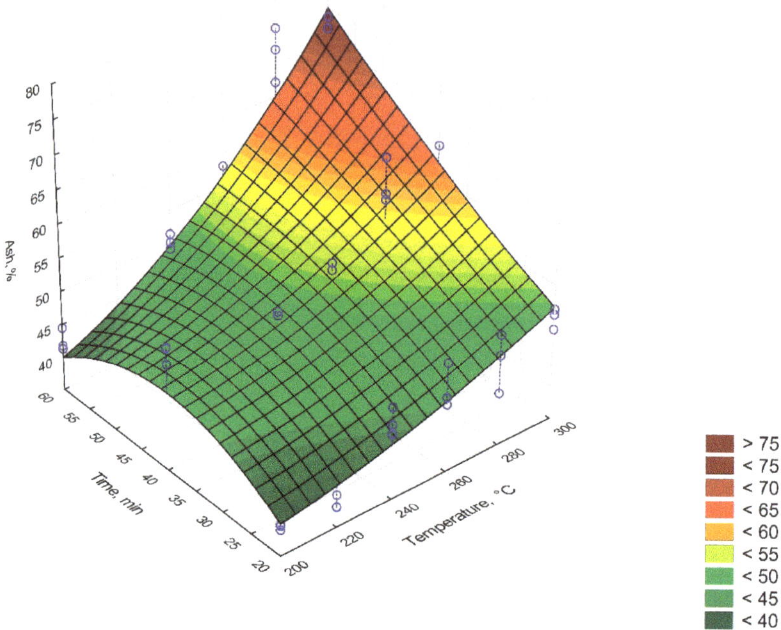

Figure 7. The influence of temperature and time on the ash content in biocoal from elephant dung.

Table 5. Statistical evaluation of ash content of biocoal from elephant dung.

Intercept/ Coefficient	Value of Intercept/ Coefficient	Standard Error	p	Lower Limit of Confidence	Upper Limit of Confidence
a_1	-2.82×10^0	6.34×10^1	0.00	-1.30×10^2	1.25×10^2
a_2	2.76×10^{-2}	4.19×10^{-1}	0.00	-8.15×10^{-1}	8.70×10^{-1}
a_3	4.22×10^{-4}	6.95×10^{-4}	0.00	-9.76×10^{-4}	1.82×10^{-3}
a_4	3.32×10^0	1.80×10^0	0.00	-3.01×10^{-1}	6.94×10^0
a_5	-3.24×10^{-2}	1.17×10^{-2}	0.00	-5.58×10^{-2}	-8.93×10^{-3}
a_6	-9.87×10^{-3}	7.12×10^{-3}	0.00	-2.42×10^{-2}	4.44×10^{-3}
a_7	3.91×10^{-7}	0.00×10^0	0.00	3.91×10^{-7}	3.91×10^{-7}

$Ash = a_1 + a_2 \cdot T + a_3 \cdot T^2 + a_4 \cdot t + a_5 \cdot t^2 + a_6 \cdot T \cdot t + a_7 \cdot T^2 \cdot t$, $R^2 = 0.83$, $R = 0.91$; T^* ranged from 200 °C to 300 °C, t^* ranged from 20 min to 60 min; * more information in Section 2.2.

The content of combustible parts (CP) decreased with time and the rise of the process temperature. Raw elephant dung had a $CP = 48.9\%$ (Table A1). During the torrefaction, the CP decreased to 28.6% at 60 min and 300 °C (Table A1, Figure 8). The analysis of variance showed numerous statistically significant differences, the majority of which occurred between 260 °C, 280 °C, and 300 °C (Table A5, Figure A3). All regression coefficients were statistically significant ($p < 0.05$) for the CP model ($R^2 = 0.67$) (Table 6).

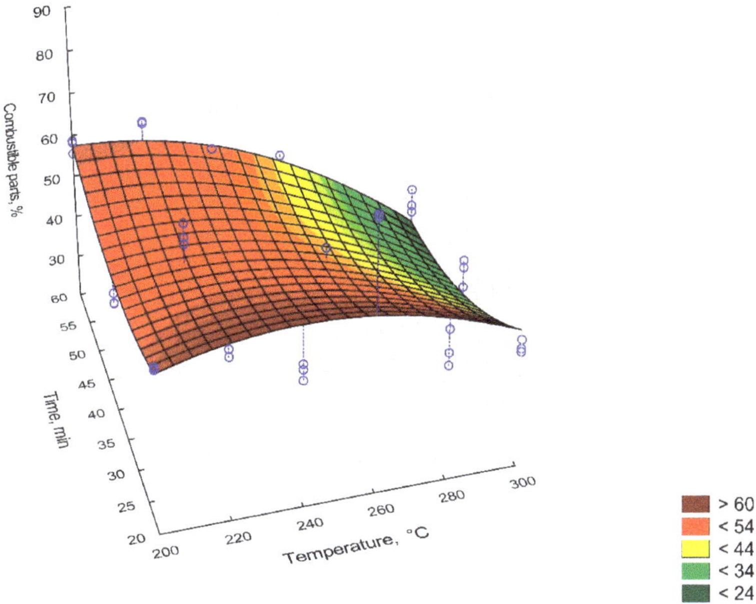

Figure 8. The influence of temperature and time on the combustible parts in biocoal from elephant dung.

Table 6. Statistical evaluation of ash content of biocoal from elephant dung.

Intercept/ Coefficient	Value of Intercept/ Coefficient	Standard Error	p	Lower Limit of Confidence	Upper Limit of Confidence
a_1	-1.19×10^2	1.14×10^2	0.00	-3.48×10^2	1.11×10^2
a_2	1.61×10^0	7.54×10^{-1}	0.00	9.02×10^{-2}	3.12×10^0
a_3	-3.10×10^{-3}	1.25×10^{-3}	0.00	-5.62×10^{-3}	-5.79×10^{-4}
a_4	-2.38×10^{-1}	3.24×10^0	0.00	-6.75×10^0	6.28×10^0
a_5	1.73×10^{-2}	2.10×10^{-2}	0.00	-2.49×10^{-2}	5.95×10^{-2}
a_6	-5.28×10^{-3}	1.28×10^{-2}	0.00	-3.11×10^{-2}	2.05×10^{-2}
a_7	-4.29×10^{-8}	0.00×10^0	0.00	-4.29×10^{-8}	-4.29×10^{-9}

$CP = a_1 + a_2 \cdot T + a_3 \cdot T^2 + a_4 \cdot t + a_5 \cdot t^2 + a_6 \cdot T \cdot t + a_7 \cdot T^2 \cdot t$, $R^2 = 0.67$, $R = 0.82$; T^* ranged from 200 °C to 300 °C, t^* ranged from 20 min to 60 min; * more information in Section 2.2.

The decrease in the *HHV* of the biocoals produced from the elephant dung was observed along with the increase of temperature and time (Figure 9, Table A1, Figure A4). The highest *HHV* was obtained for the biocoal generated at 200 °C and 60 min. A similar trend was discovered by Li et al., [33]. They explained this phenomenon by the effect of specific biocoal properties (pH; C, H, N, S, O content; specific surface area) and noticed also a possibility of predicting the biocoal yield of a group of feedstocks with similar physiochemical properties.

The average *HHV* was 13 MJ·kg^{-1} and was higher than the *HHV* of raw elephant dung (by 1.59 MJ·kg^{-1}) and higher than the lowest *HHV* for the biocoal obtained at 300 °C and 60 min (by 6.51 MJ·kg^{-1}). The *HHV* is affected by the high ash content in the biocoals and raw material. Thus, it was decided to estimate the value of *HHV* on an ash-free basis (*HHV*$_{daf}$). The highest average *HHV*$_{daf}$ was obtained for the biocoal generated at 280 °C and for 60 min (27.20 MJ·kg^{-1}) (Figure 10, Table A1). Regression coefficients for the *HHV* and *HHV*$_{daf}$ were statistically significant ($p < 0.05$), the proposed model worked well for *HHV* but was less representative for *HHV*$_{daf}$ (R^2 were 0.74 and 0.21, respectively) (Tables 7 and 8). Analysis of the variance of average values of *HHV* showed statistically significant differences between the results for 280 °C and 300 °C and 40 & 60 min ($p < 0.05$) (Figure A4, Table A6). This result has practical implications for the collection and initial processing of elephant dung to minimize mineral ash content and impurities and to maximize the *HHV*.

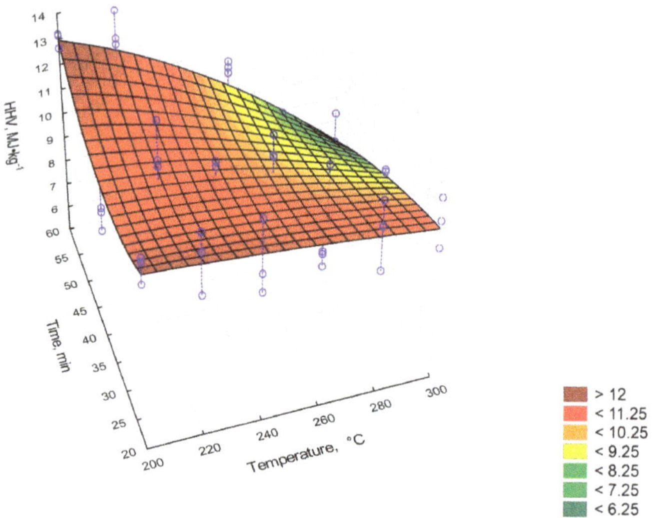

Figure 9. The influence of temperature and time on the high heating value *(HHV)* in biocoal from elephant dung.

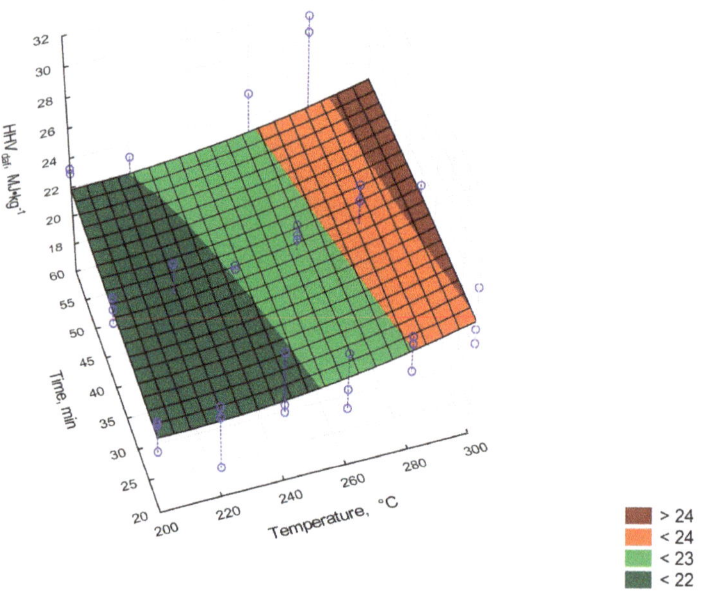

Figure 10. The influence of temperature and time on the HHV_{daf} in biocoal from elephant dung.

Table 7. Statistical evaluation of the high heating value of biocoal from elephant dung.

Intercept/ Coefficient	Value of Intercept/ Coefficient	Standard Error	p	Lower Limit of Confidence	Upper Limit of Confidence
a_1	2.25×10^1	1.50×10^1	0.00	-7.67×10^0	5.27×10^1
a_2	-3.04×10^{-2}	9.92×10^{-2}	0.00	-2.30×10^{-1}	1.69×10^{-1}
a_3	-2.45×10^{-6}	1.65×10^{-4}	0.00	-3.34×10^{-4}	3.29×10^{-4}
a_4	-6.30×10^{-1}	4.26×10^{-1}	0.00	-1.49×10^0	2.27×10^{-1}
a_5	6.68×10^{-3}	2.76×10^{-3}	0.00	1.12×10^{-3}	1.22×10^{-2}
a_6	1.84×10^{-3}	1.69×10^{-3}	0.00	-1.55×10^{-3}	5.23×10^{-3}
a_7	-8.13×10^{-8}	0.00×10^0	0.00	-8.13×10^{-8}	-8.13×10^{-8}

$HHV = a_1 + a_2 \cdot T + a_3 \cdot T^2 + a_4 \cdot t + a_5 \cdot t^2 + a_6 \cdot T \cdot t + a_7 \cdot T^2 \cdot t$, $R^2 = 0.74$, $R = 0.86$; T^* ranged from 200 °C to 300 °C, t^* ranged from 20 min to 60 min; * more information in Section 2.2.

Table 8. Statistical evaluation of high heating value on the dry ash-free basis of biocoal from elephant dung.

Intercept/ Coefficient	Value of Intercept/ Coefficient	Standard Error	p	Lower Limit of Confidence	Upper Limit of Confidence
a_1	3.54×10^1	2.99×10^1	0.00	-2.48×10^1	9.56×10^1
a_2	-1.28×10^{-1}	1.98×10^{-1}	0.00	-5.25×10^{-1}	2.70×10^{-1}
a_3	2.91×10^{-4}	3.28×10^{-4}	0.00	-3.69×10^{-4}	9.52×10^{-4}
a_4	-1.04×10^{-1}	8.49×10^{-1}	0.00	-1.81×10^0	1.60×10^0
a_5	6.68×10^{-4}	5.50×10^{-3}	0.00	-1.04×10^{-2}	1.17×10^{-2}
a_6	4.75×10^{-4}	3.36×10^{-3}	0.00	-6.29×10^{-3}	7.23×10^{-3}
a_7	-1.03×10^{-8}	0.00×10^0	0.00	-1.03×10^{-8}	-1.03×10^{-8}

$HHV_{daf} = a_1 + a_2 \cdot T + a_3 \cdot T^2 + a_4 \cdot t + a_5 \cdot t^2 + a_6 \cdot T \cdot t + a_7 \cdot T^2 \cdot t$, $R^2 = 0.21$, $R = 0.45$; T^* ranged from 200 °C to 300 °C, t^* ranged from 20min to 60 min; * more information in Section 2.2.

3.3. Result of the Thermogravimetric Analysis (TGA) of Elephant Dung

Table 9 summarizes kinetics parameters based on the TGA analyses and the mass loss data.

Table 9. The values of reaction rate constants and activation energy for elephant dung torrefaction.

T, °C	T^{-1}, °C^{-1}	k, s^{-1}	$\ln(k)$, s^{-1}	Ea, J·mol^{-1}
200	2.11×10^{-3}	1.16×10^{-5} [a]	−11.40	
220	2.03×10^{-3}	1.24×10^{-5} [a]	−11.30	
240	1.95×10^{-3}	1.49×10^{-5} [a]	−11.10	17,700
260	1.88×10^{-3}	1.50×10^{-5} [a]	−11.10	
280	1.81×10^{-3}	1.92×10^{-5} [a,b]	−10.90	
300	1.75×10^{-3}	2.73×10^{-5} [b]	−10.50	

a, b—letters present a lack of statistically significant differences between k values ($p < 0.05$).

The obtained values of k were analyzed by ANOVA, which showed that there were statistically significant differences ($p < 0.05$) for biocoal produced at 300 °C, and those obtained at 200 °C, 220 °C, 240 °C, and 260 °C, respectively. There were no statistical differences between k for 280 and 300 °C and k for 200–260 °C range. Kim et al. indicated that different optimal temperatures should be selected for different types of manure to maximize the energetic retention efficiency [34]. The energy yield of hydrochar (48.0–71.9%) is higher than that of pyrolysis char (31.5–52.4%), implying that the carbonization process, rather than the reaction temperature, is also a key factor that affects the energy yield of manure [35]. The TGA analysis showed the most substantial mass decrease in the first repetition to 54% of the initial mass of the sample, while in the second and third repetitions, the mass decreased to 64% and 62%. The loss of mass began at a temperature of ~300 °C, and it started to stabilize after exceeding ~600 °C (Figure 11).

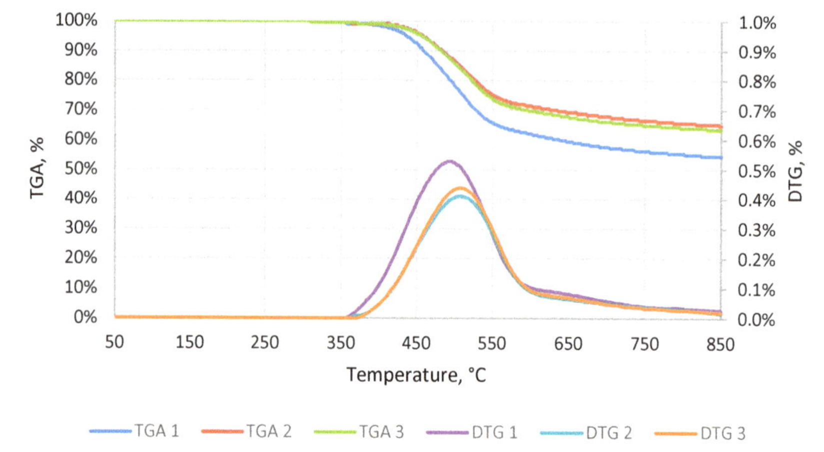

Figure 11. The thermogravimetric characteristic of elephant dung.

New knowledge on the substrates of elephant dung was gained from the TGA analyzes. There was a characteristic peak start at ~330 °C with a maximum at ~500 °C, most likely related to the decomposition of undigested (by elephant) cellulose and lignin from consumed biomass. The decomposition of cellulose and lignin takes place at 305–375 °C and 250–500 °C, respectively [36]. No degradation of hemicellulose was observed based on the DTG (derivative thermogravimetry) analysis. The decomposition of hemicellulose takes place at 225–325 °C [36]. However, the apparent lack of mass change in this temperature range (Figure 11) does not necessarily indicate a lack of hemicelluloses content. It is also likely that particular decompositions could be superimposed [36] and could not be detected by the lack of precision of the used thermogravimetric analyzer.

3.4. Differential Scanning Calorimetry (DSC) of Elephant Dung

DSC analysis showed that during heating, two endoenergetic transformations occurred (Figure 12).

At the beginning of the experiment, the energy was supplied to the sample to raise the temperature of the system. The first observation was that transformation began at 37 °C. Here, the energy was delivered to heat a sample and to initiate its transformation, which reached its maximum value at 80 °C and ended at 146 °C. The total energy demand for this first transformation was 66.17 J·g^{-1}. After the first transformation ended, the energy needed only for heating the sample was supplied to the system (146–158 °C). The second transformation began at 158 °C, reached its maximum at 216 °C, and ended at 252 °C, requiring only 9.76 J·g^{-1}. After the second transformation occurred, the energy required for heating decreased significantly. After T > 252 °C the exothermic reaction occurred.

The total energy demand for the whole process including heating and transformations of dry elephant dung was 485.37 kJ·kg^{-1} for the −20 to 300 °C range. The estimate for process energy demand calculated by model for torrefaction [30] decreased to 484.81 kJ·kg^{-1}, and it was due to mass loss during the process. In addition, the heating and evaporation of the water contained in raw elephant dung (moisture content 49.19%), results in the additional 1275.49 kJ·kg^{-1} (Equation (8)) energy demand. Thus, the total energy demand for processing of raw elephant dung (heating, moisture evaporation, and torrefaction) is 1760.30 kJ·kg^{-1}.

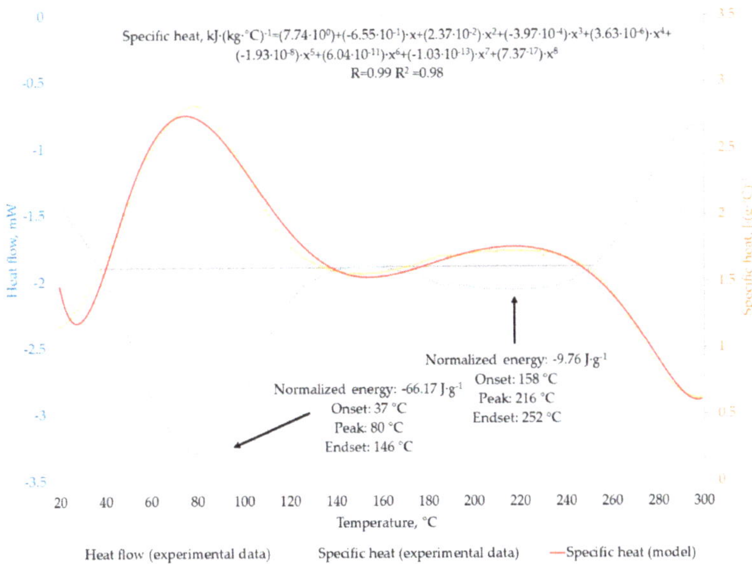

Figure 12. The differential scanning calorimetry (DSC) characteristic of elephant dung.

4. Discussion

4.1. The Impact of Technological Parameters on the Efficiency of the Process

A related torrefaction study carried out on cow manure showed that the *MY* of torrefaction decreased with the increase of the process temperature [37], similar to the finding in this research. The torrefied elephant dung (200–300 °C at 40 min) had the *MY* of 100–68%, whereas it was 90–55% for cow manure at the same process conditions [37]. Differences in *MY* could be explained by a greater decomposition of biodegradable substrates at lower temperatures. Also, elephant dung had higher moisture and *OM* content compared with the cow manure. In addition, it has been reported that it is possible to change specific surface area (SSA) as a result of morphological changes due to thermal condensation, and it could be exploited in different materials [38]. The energy yield of torrefaction of cow manure decreased from around 92% at 200 °C to approximately 57% at 300 °C, whereas elephant dung were of 110% and 60%, respectively. The *EDr* ratio for cow manure had the same downtrend as elephant dung [37]. It was also noticed that there are different degradation processes in the studied range of 200–300 °C. Lignocellulose degradation occurs at approximately 120 °C; hemicellulose degradation occurs at 200–260 °C; cellulose degradation occurs at 240–350 °C; while lignin degradation occurs at 280–350 °C [39], which due to the observation of narrow temperature ranges could have affected the lack of a decrease or increase trend in the case of obtained moisture and *MY*.

4.2. Proximate Analyses of Elephant Dung and Biocoals

The average moisture content in the elephant dung was 49.19%. The moisture content of dung depends on the amount of water consumed by the animal. For example, pig manure could have a moisture content of ~35–82%, whereas cow manure is of ~66–97% [40–42]. In the case of poultry manure, moisture content ranges from ~5 to 40% [40]. The *OM* content in the studied elephant dung was 48.09% (d.m.). For comparison, the *OM* content for Indian elephant and rhinoceros were 52% and 56%, respectively [43]. For yet another case of the cattle manure, an *OM* content was ~74% [44].

These *OMs* are much lower than those reported in related torrefaction studies for pruned biomass of Paulownia (90%) [45], or brewery spent grain (96%) [46].

The elephant dung had a higher ash content (50.81%, d.m.) than the maximum content of ash in pig manure (21.4% d.m.), cow manure (32.8% d.m.) and chicken manure (34% d.m.) [40,47]. The *HHV* of elephant dung was 11.41 MJ·kg^{-1} and was lower than *HHV* in chicken manure (13 MJ·kg^{-1}), cow manure (12.7–17.2 MJ·kg^{-1}), or pig manure (18.1–19.5 MJ·kg^{-1}) [37,40,41,48]. The low value of *HHV* is likely caused by high ash content, i.e., the calculated ash-free *HHV* was as high as 23 MJ·kg^{-1}.

The *HHV* of the torrefied dung was not much higher than the raw sample (Table A1). For biocoal, the highest *HHV* was 13 MJ·kg^{-1} (260 °C, 60 min), and a further increase in temperature and time caused a decrease in its value. The low increase of *HHV* in comparison to the raw base for cow dung was reported by Pahla et al. [37] and *HHV* increased from 16.78 to 18.64 MJ·kg^{-1} (at 300 °C). A small increase of *HHV* in dung biocoal is directly affected by a low amount of fixed carbon (high amount of ash content). During torrefaction, fixed carbon is enhanced by thermal degradation of hemicellulose and part of cellulose and lignin [49]. The decomposition of these constituents results in releases of compounds with low energy content, leaving organic compounds with higher energy content [50]. Cow manure, similarly to elephant dung did not experience high *HHV* enhancement likely because it had less *OM* and more ash content. Pulka et al. [28] tested sewage sludge via torrefaction and met the same problem—the highest value of *HHV* for biocoal generated at 260 °C, 60 min, and further temperature increase decreased *HHV*. Therefore, it may be assumed that at a temperature > 260 and time > 60 min, some organic components from elephant dung and sewage sludge start to decompose and release volatiles with higher energy content.

There was no observed relationship between the moisture content and the process temperature and time for the biocoals from elephant dung. This is likely because dry material was used for the torrefaction process. Small differences in the moisture content of biocoals can result from the time between their generation and the determination of the moisture content experiment. Stored biocoals can adsorb moisture (e.g., from the air), making biomass-derived fuels less advantageous compared with coal [50].

There was a sharp drop in the *OM* and the simultaneous increase in ash content for torrefaction above 260 °C. This also caused a decrease of *HHV* and an increase in the *HHV*$_{daf,}$ especially in the biocoals produced at 260 °C and 300 °C. A practical implication is that the torrefaction process conducted at temperatures from 200 °C to 260 °C (regardless of time) will have a small impact on the decrease of *HHV* of biocoals.

Furthermore, it could be recommended that torrefaction at 200 °C for 20 min (lowest temperature and shortest time) is needed for the maximization of the *HHV* and minimization of the cost of the torrefaction process. In addition, a lack of significant differences ($p < 0.05$) in 200–260 °C allows us to use torrefaction of elephant dung as a low-tech technology, i.e., one that can be controlled without an accurate measurement system. It is especially important for rural areas. Also, during torrefaction of a more substantial amount of the dung, it would be challenging to evenly heat and then cool down fast all the processed material. However, based on the apparent lack of effect in this research, the risk of generating substandard biocoals appears to be relatively low.

The highest *HHV*$_{daf}$ value (27.2·MJ·kg^{-1}) was observed for 280 °C, 60 min (Table A1). This value is theoretical, and it is worth considering ways of reducing the ash content in the elephant dung, because it may have a high energetic potential after processing. Considering ash-free elephant dung after torrefaction, it is possible to obtain better solid fuel than commercially-available pellets. For example, pellets made from pine sawdust, wheat straw, corn settlements, agricultural residues have *HHV* of 19.5, 17.5, 18.8, 18.1 MJ·kg^{-1}, (*HHV*$_{daf}$ 19,6, 19.0, 19.0, 19.8 MJ·kg^{-1}) (Table A8) respectively [51]. These values are still relatively low when compared to ash-free biocoal from elephant dung of 27.2 MJ·kg^{-1}.

The ash in elephant dung is derived from two primary sources, (1) ash introduced during collecting, transporting, storing, and processing, (2) biogenic ash inside plant tissue consumed by an elephant. The sum of these sources is referred to as ash content. Biogenic ash could be removed from biomass using air separation. For woody pine forest residue, air separation costs ~2.23 $·Mg^{-1} of biomass to reduce 40% of total biogenic ash to <7% of total biomass [52]. Ash could also be removed from biomass cells via chemical pre-processing that solubilize it. Here, knowledge of the exact morphology and chemical state of the ash is needed to determine the most effective removal methods [52]. From a practical point of view, elephant dung should be collected with the least soil impurities as possible. Next, during transportation, drying, etc. the dung should not be exposed to dust. If prevention is not enough, air separation could be considered, due to its relatively low operational cost. Nevertheless, dung morphology is important factor for air separation. Dung is much more brittle and lighter than wood. Because of this, chipped particles of dung could be lighter than mineral impurities causing the different share of ash in particular fractions than in the case of wood. Although some chemical pre-processing technologies have a high level of ash removal (over 90% removal of alkaline earth and alkali metals) [52], their technological infrastructure and cost would be difficult to adopt in underserved areas.

Another important aspect is the issues related to the supply chain, which may influence the quality of biocoal and efficiency of the process. The collection of elephant dung has a dispersed character with a random accumulation ratio in one specific localization, especially when elephants live in natural habitats. The dung usually is collected directly from the ground, which may increase the ash content. However, when dung is exposed to climatic conditions (especially to wind and sun), the overall effect might be beneficial to drying, which brings benefits related to transportation and torrefaction efficiency. Pre-dried material is more suitable for collection, transportation (less water to be transported), and is less prone to decay. In the case of breeding of elephants or using them as work animals (as practiced in South-East Asia), the accumulation of dung in one specific area is more likely. Natural drying maybe not be sufficient. Therefore, one solution could be pre-drying in the dedicated dryer, which could use a warm air stream for water removal. Solar energy could be used as a heat source. Such solution could solve several practical problems: i) the long-range transport of untreated and wet dung to processing sites that is energy inefficient, while a significant portion of the transportation costs are being used to transport water [5]; ii) the long-term storage of raw biomass can be problematic and impractical because the piled biomass can decompose over time resulting in the decrease of useful *HHV* [7].

4.3. Thermogravimetric Analysis of Raw Material and Kinetic Parameters of Torrefaction

Reported TGA analyses of elephant dung are the first of their kind in the literature. A comparison of kinetic parameters with the literature is then confounded because of the variety of determination methods used for other materials. For this reason, we discuss the kinetics of a subset of the most common and related substrates. We considered the elephant diet consisting mostly of grasses, and the activation energy for some grass plants is available. The activation energy of wheat straw and sorghum determined for the 250–450 °C range was 176 kJ·mol^{-1} and kJ·mol^{-1}, respectively [53]. For comparison, lignocellulose materials (eg., woody biomass) have an E_a of 103–165 kJ·mol^{-1} [54,55]. The values presented in this paper were obtained for non-isothermal conditions and pyrolysis temperature range.

In this work, the E_a and the reaction rate constants were determined in isothermal conditions and a temperature range of 200–300 °C. The same conditions and temperatures were used previously by Pulka et al. [56], who tested sewage sludge (SS) with high ash content, and Syguła et al. [57] who tested spent mushroom compost (MSC). The E_a for torrefaction process of elephant dung was 18 kJ·mol^{-1}, and k values were increasing with process temperature from 1.16×10^{-5}·s^{-1} to 2.73×10^{-5}·s^{-1} (from 200 to 300 °C), respectively. In the case of SS, the E_a was 12 kJ·mol^{-1}, and the k value increased from 4.02×10^{-5}·s^{-1} to 6.71×10^{-5}·s^{-1} (from 200 to 300 °C), respectively [56]. In the case of MSC, the E_a was 22.2 kJ·mol^{-1}, and the k value increased from 1.7×10^{-5}·s^{-1} to 4.6×10^{-5}·s^{-1} (from 200 to 300 °C), respectively [57]. Differences could be a result of biomass origin, and organic matter content. *OM* in SS was 56% d.m. [56], 76% d.m. in MSC [57], and 50% d.m. in elephant dung—Table A1).

It should also be noted that the greatest E_a was determined for MSC, which had the highest *OM* content, and much smaller during the torrefaction of elephant dung and SS, where *OM* contents were lower by ~20%. An opposite trend was observed in the case of the *k* value, which was the highest during the torrefaction of SS, followed by MSC and elephant dung. This may indicate that the content of *OM* is one of the critical drivers of the waste's kinetic properties, such as the *Ea* and possibly *k*.

4.4. Differential Scanning Calorimetry of Raw Material

DSC analysis showed that two endothermic reactions (37–146 °C and 158–252 °C) and one exothermic reaction (252–300 °C) occur during the torrefaction process (Figure 12). The first transformation observed on DSC plot may be attributed to water evaporation. Interestingly, the elephant dung was dried at 105 °C before the DSC test. Thus, the presence of water in a previously dried sample could be due to the hygroscopicity (the sample absorbed some water from the atmosphere before the test; i.e., biocoals are known to be affected by this phenomenon) [58]. The first transformation ended above drying temperature (105 °C), so it is probably associated with bound water evaporation. The nature of the second endothermic transformation is unknown. To our knowledge, there are no DSC data of elephant dung to compare. This transformation may be related to residue hemicellulose degradation. Degradation of hemicellulose takes place at a lower temperature range (225–325 °C) than the degradation of cellulose (305–375 °C) [36]. After the second endothermic transformation ended, the heat flow starts to decrease, which is related to an exothermic reaction (253–300 °C). This exothermic reaction corresponds to mass loss observed on TG/DTG plot observed at the beginning of the process (Figure 11). Interestingly, neither of the endothermic reactions were apparerent in the TG/DTG plot (Figure 11). This might be a result of insufficient precision in the use of the laboratory balance, or due to transformations that were not related to mass loss. In general, endothermal reactions are related to depolymerization and volatilization process, whereas exothermic transformations are due to the charring process [59] phenomenon, the DSC plot shows that the elephant dung torrefaction is an (overall) endothermic process and it requires energy delivery. Some energy cost savings might be realized by using the torrefied elephant dung as a fuel for the torrefaction process (Figure 1).

High ash content 50.81% (Table A1) is not without significance. It makes measurements of TGA and DSC less accurate because smaller mass loses in organic compounds were measured. In the case of DSC, the endothermic reactions of <200 °C that were found could also be associated with water evaporation from components of ash such as chlorine and potassium [60]. The growth of the mineral fraction lowers the activation energy of the pyrolysis reaction, and accelerates exothermic thermochemical conversion reactions [61].

5. Conclusions

Initial valorization of elephant dung by torrefaction is proposed as a possible low-tech fuel production in rural areas with abundant supply. Proposed valorization could be used in households for cooking and heating. These studies have expanded knowledge on the possibilities of torrefaction of elephant dung and provided practical knowledge about the fuel properties of torrefied elephant dung, as high heating value, combustible parts, ash content, and organic matter content. Based on the results, models of torrefaction of elephant dung with kinetics parameter evaluation have been proposed. The following conclusions arise from this research:

- Torrefaction improves the higher heating value of elephant dung. The torrefied elephant dung has an $HHV = 13$ MJ·kg^{-1} compared to the $HHV = 11.41$ MJ·kg^{-1} for unprocessed dung.
- Minimal process controls appear to be needed, and thus, scaling the torrefaction up to larger batches of dung is feasible, but due to lack of data, these options need more tests on a technical scale. Biocoals with similar quality are obtained for 200 °C to 260 °C range regardless of the duration of the process (20 to 60 min).
- The recommended temperature of the torrefaction for elephant dung is 200 °C, due to the lack of significant improvements in fuel properties with increasing process temperature.

- The activation energy for torrefaction of elephant dung at 200~300 °C was 17.7 J·mol^{-1}, and the reaction rate constant increased from 1.16×10^{-5}·s^{-1} to 2.73×10^{-5}·s^{-1}.

- The total energy needed to heat the dry elephant dung from 20 °C to 300 °C was approximately 485 kJ·kg^{-1} (obtained in laboratory conditions), and 484.81 kJ·kg^{-1} (obtained from calculations) after the mass loss during the process is factored in. The total energy demand for drying and torrefaction was the total amount of energy for processing (heating, moisture evaporation, and torrefaction) was 1760.30 kJ·kg^{-1}.

This research has shown that there is a potential in using elephant dung as a substrate for torrefaction and its valorization as an improved fuel source. The next step should be to identify the technological parameters for the torrefaction of elephant dung. This is important for investment analysis and technology design, particularly in rural areas.

Author Contributions: Conceptualization, P.S., M.H. and K.Ś.; methodology, M.H.; software, P.S., S.K., and K.Ś.; validation, P.S., K.Ś., and M.H.; formal analysis, M.H.; investigation, M.H. and S.K.; resources, M.H. and A.B.; data curation, P.S., K.Ś., A.B. and J.A.K.; writing—original draft preparation, P.S., M.H. and K.Ś.; writing—review and editing, P.S., K.Ś., S.S.-D., J.A.K.; visualization, P.S. and K.Ś.; supervision, A.B., J.A.K., and P.M.; project administration, P.S.; funding acquisition, P.S., A.B., and J.A.K.

Funding: The research was funded by the Polish Ministry of Science and Higher Education (2015–2019), the Diamond Grant program # 0077/DIA/2015/14. "The PROM Programme - International scholarship exchange of Ph.D. candidates and academic staff" is co-financed by the European Social Fund under the Knowledge Education Development Operational Programme PPI/PRO/2018/1/00004/U/001. The authors would like to thank the Fulbright Foundation for funding the project titled "Research on pollutants emission from Carbonized Refuse Derived Fuel into the environment", completed at Iowa State University. In addition, this project was partially supported by the Iowa Agriculture and Home Economics Experiment Station, Ames, Iowa. Project no. IOW05556 (Future Challenges in Animal Production Systems: Seeking Solutions through Focused Facilitation) sponsored by Hatch Act and State of Iowa funds.

Conflicts of Interest: The authors declare no conflict of interest.

Appendix A

Table A1. Summary of proximate analysis of the tested elephant dung and biocoals resulting from its torrefaction.

Sample		Moisture, %	Organic Matter Content, %	Ash, %	HHV, MJ·kg^{-1}	HHV$_{daf}$, MJ·kg^{-1}
Elephant dung		49.19 ± 5.84	48.90 ± 5.79	50.81 ± 5.84	11.41 ± 1.34	23.18 ± 2.39
200 °C	20 min	3.33 ± 0.08	60.44 ± 0.46	39.37 ± 0.44	12.75 ± 0.58	21.75 ± 1.15
	40 min	1.14 ± 0.02	47.50 ± 1.42	52.40 ± 1.42	10.14 ± 0.51	21.56 ± 0.87
	60 min	2.35 ± 0.08	57.35 ± 1.69	42.51 ± 1.70	13.00 ± 0.31	23.16 ± 0.19
220 °C	20 min	2.11 ± 0.15	61.23 ± 1.04	38.65 ± 1.05	12.47 ± 1.31	20.77 ± 2.30
	40 min	2.15 ± 0.05	60.22 ± 2.52	39.77 ± 2.50	12.34 ± 1.01	21.00 ± 2.48
	60 min	1.90 ± 0.06	60.21 ± 0.27	39.76 ± 0.24	12.82 ± 0.72	21.70 ± 1.24
240 °C	20 min	2.11 ± 0.12	53.57 ± 2.09	46.50 ± 2.01	11.80 ± 1.56	22.48 ± 2.24
	40 min	1.03 ± 0.04	49.91 ± 1.12	50.03 ± 1.08	10.74 ± 0.79	21.71 ± 1.27
	60 min	0.96 ± 0.05	49.79 ± 1.11	50.11 ± 1.13	9.51 ± 0.50	19.24 ± 0.59
260 °C	20 min	3.20 ± 0.06	52.96 ± 3.14	47.52 ± 3.30	11.39 ± 0.33	21.79 ±1.93
	40 min	0.88 ± 0.10	47.63 ± 2.92	52.24 ± 2.97	11.25 ± 0.50	23.77 ± 0.51
	60 min	1.07 ± 0.04	44.82 ± 2.58	55.07 ± 2.62	10.34 ± 0.24	23.33 ± 1.93
280 °C	20 min	2.23 ± 0.15	52.21 ± 4.41	47.60 ± 4.45	11.80 ± 1.45	23.00 ± 1.25
	40 min	2.85 ± 0.26	37.23 ± 3.26	62.59 ± 3.31	8.66 ± 1.22	23.87 ± 2.92
	60 min	1.64 ± 0.26	28.26 ± 3.97	71.48 ± 3.99	7.54 ± 0.32	27.20 ± 3.57
300 °C	20 min	2.61 ± 0.25	49.47 ± 1.47	50.21 ± 1.53	11.64 ± 1.02	24.01 ± 1.99
	40 min	1.99 ± 0.26	39.09 ± 3.47	60.89 ± 3.46	9.05 ± 0.32	23.69 ± 1.25
	60 min	1.24 ± 0.14	28.66 ± 2.92	71.25 ± 2.92	6.49 ± 0.71	22.86 ± 0.79

Table A2. Values of mass yield, energy yield, and energy densification ratio for biocoals.

Sample		Mass Yield, %	Energy Yield, %	Energy Densification Ratio, %
	20 min	91.42	102.11	1.12
200 °C	40 min	98.65	107.78	1.09
	60 min	95.59	108.91	1.13
	20 min	95.43	104.25	1.09
220 °C	40 min	93.16	100.74	1.08
	60 min	90.43	101.62	1.12
	20 min	98.12	101.43	1.03
240 °C	40 min	92.78	87.33	0.94
	60 min	89.36	74.46	0.83
	20 min	97.07	95.66	0.99
260 °C	40 min	88.63	87.34	0.99
	60 min	90.01	81.55	0.90
	20 min	71.83	68.89	0.96
280 °C	40 min	53.21	46.29	0.87
	60 min	63.33	51.59	0.81
	20 min	63.28	58.28	0.92
300 °C	40 min	66.58	56.50	0.85
	60 min	73.18	62.58	0.86

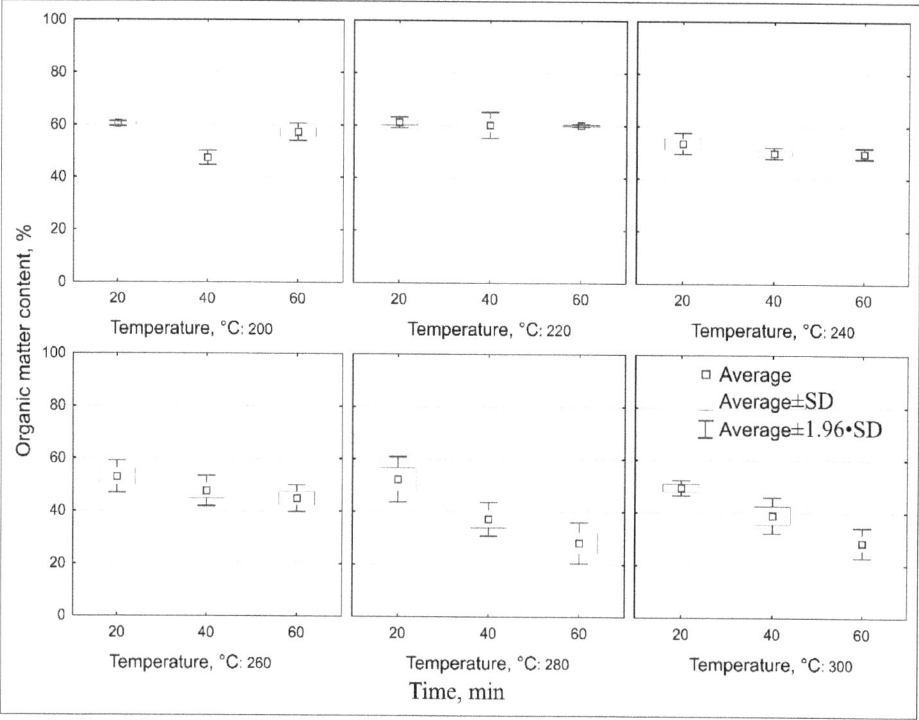

Figure A1. Presentation of differences in individual groups (of torrefaction time) for organic matter content in biocoals from elephant dung.

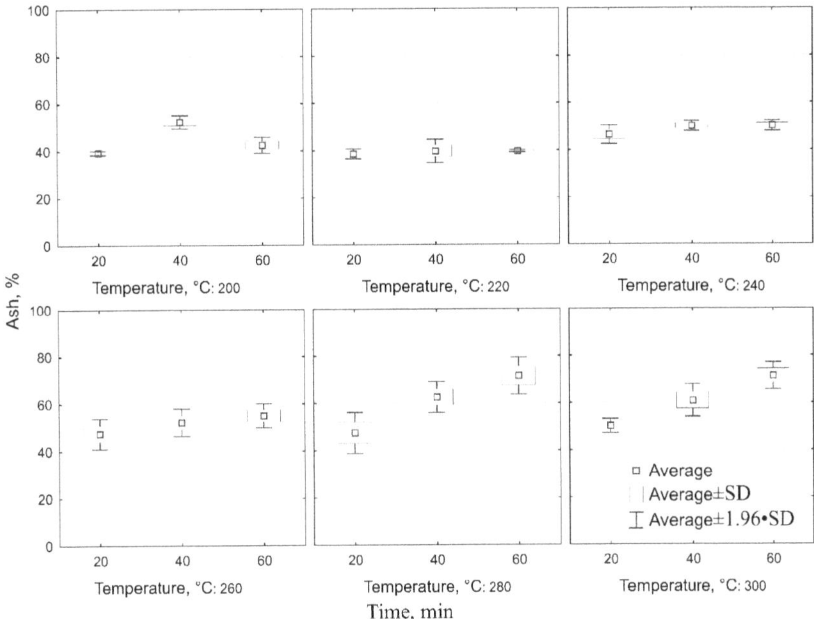

Figure A2. Presentation of differences in individual groups (of torrefaction time) for ash content in biocoals from elephant dung.

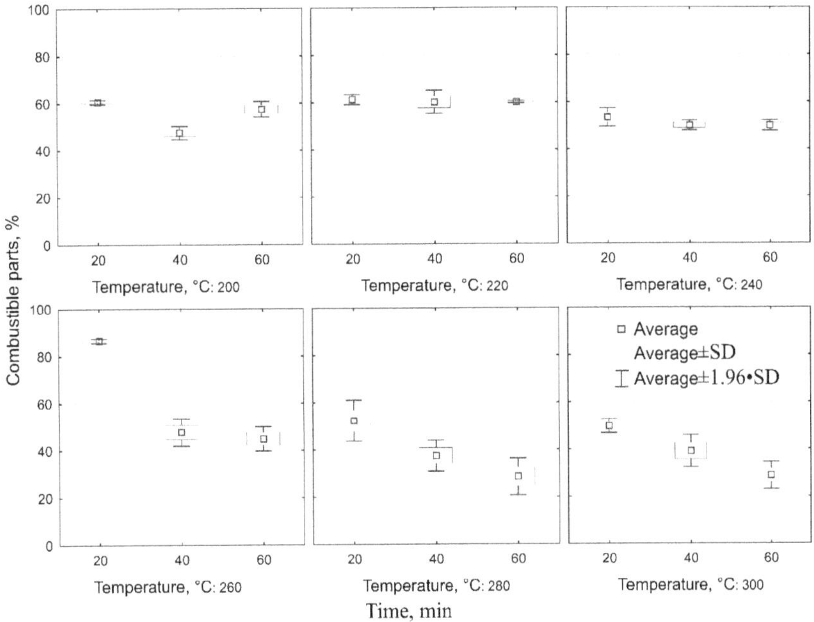

Figure A3. Presentation of differences in individual groups (of torrefaction time) for combustible parts in biocoals from elephant dung.

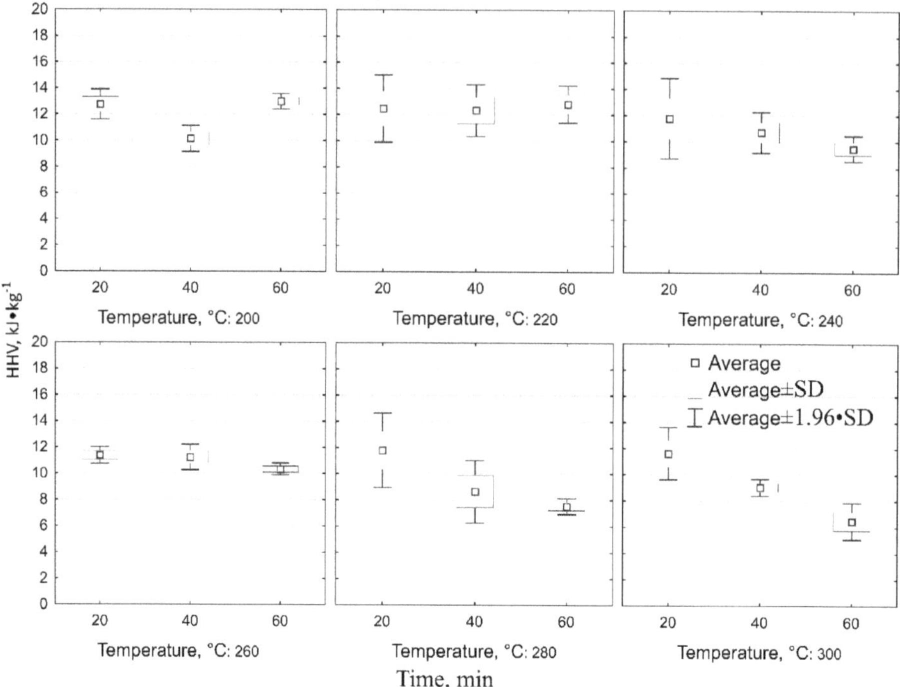

Figure A4. Presentation of differences in individual groups (of torrefaction time) for the high heating value of biocoals from elephant dung.

Table A3. Analysis of variance for organic matter (OM) content.

Tukey Test for OM; a Bold Font Signifies Statistically Significant Difference ($p < 0.05$)		200	200	200	220	220	220	240	240	240	260	260	260	280	280	280	300	300	300
		20	40	60	20	40	60	20	40	60	20	40	60	20	40	60	20	40	60
200	20		**0.00**	0.98	1.00	1.00	1.00	0.12	**0.00**	**0.00**	0.06	**0.00**	**0.00**	0.03	**0.00**	**0.00**	**0.00**	**0.00**	**0.00**
200	40	**0.00**		**0.00**	**0.00**	**0.00**	**0.00**	0.27	1.00	1.00	0.43	1.00	1.00	0.68	**0.00**	**0.00**	1.00	**0.02**	**0.00**
200	60	0.98	**0.00**		0.90	0.99	0.99	0.91	0.07	0.06	0.77	**0.00**	**0.00**	0.53	**0.00**	**0.00**	**0.04**	**0.00**	**0.00**
220	20	1.00	**0.00**	0.90		1.00	1.00	0.05	**0.00**	**0.00**	**0.02**	**0.00**	**0.00**	**0.01**	**0.00**	**0.00**	**0.00**	**0.00**	**0.00**
220	40	1.00	**0.00**	0.99	1.00		1.00	0.15	**0.00**	**0.00**	0.08	**0.00**	**0.00**	0.03	**0.00**	**0.00**	**0.00**	**0.00**	**0.00**
220	60	1.00	**0.00**	0.99	1.00	1.00		0.16	**0.00**	**0.00**	0.08	**0.00**	**0.00**	0.03	**0.00**	**0.00**	**0.00**	**0.00**	**0.00**
240	20	0.12	0.27	0.91	0.05	0.15	0.16		0.93	0.91	1.00	0.30	**0.01**	1.00	**0.00**	**0.00**	0.85	**0.00**	**0.00**
240	40	**0.00**	1.00	0.07	**0.00**	**0.00**	**0.00**	0.93		1.00	0.99	1.00	0.55	1.00	**0.00**	**0.00**	1.00	**0.00**	**0.00**
240	60	**0.00**	1.00	0.06	**0.00**	**0.00**	**0.00**	0.91	1.00		0.98	1.00	0.59	1.00	**0.00**	**0.00**	1.00	**0.00**	**0.00**
260	20	0.06	0.43	0.77	**0.02**	0.08	0.08	1.00	0.99	0.98		0.47	**0.03**	1.00	**0.00**	**0.00**	0.95	**0.00**	**0.00**
260	40	**0.00**	1.00	**0.00**	**0.00**	**0.00**	**0.00**	0.30	1.00	1.00	0.47		0.99	0.72	**0.00**	**0.00**	1.00	**0.02**	**0.00**
260	60	**0.00**	1.00	**0.00**	**0.00**	**0.00**	**0.00**	**0.01**	0.55	0.59	**0.03**	0.99		0.07	0.06	**0.00**	0.70	0.35	**0.00**
280	20	**0.03**	0.68	0.53	**0.01**	**0.03**	**0.03**	1.00	1.00	1.00	1.00	0.72	0.07		**0.00**	**0.00**	1.00	**0.00**	**0.00**
280	40	**0.00**	**0.00**	**0.00**	**0.00**	**0.00**	**0.00**	**0.00**	**0.00**	**0.00**	**0.00**	**0.00**	0.06	**0.00**		**0.01**	**0.00**	1.00	**0.02**
280	60	**0.00**	**0.00**	**0.00**	**0.00**	**0.00**	**0.00**	**0.00**	**0.00**	**0.00**	**0.00**	**0.00**	**0.00**	**0.00**	**0.01**		**0.00**	**0.00**	1.00
300	20	**0.00**	1.00	**0.04**	**0.00**	**0.00**	**0.00**	0.85	1.00	1.00	0.95	1.00	0.70	1.00	**0.00**	**0.00**		**0.00**	**0.00**
300	40	**0.00**	**0.02**	**0.00**	**0.00**	**0.00**	**0.00**	**0.00**	**0.00**	**0.00**	**0.00**	**0.02**	0.35	**0.00**	1.00	**0.00**	**0.00**		**0.00**
300	60	**0.00**	**0.00**	**0.00**	**0.00**	**0.00**	**0.00**	**0.00**	**0.00**	**0.00**	**0.00**	**0.00**	**0.00**	**0.00**	**0.02**	1.00	**0.00**	**0.00**	

Table A4. Analysis of variance for ash content.

Tukey Test for Ash Content; a Bold Font Signifies Statistically Significant Difference ($p < 0.05$)		200	200	200	220	220	220	240	240	240	260	260	260	280	280	280	300	300	300
		20	40	60	20	40	60	20	40	60	20	40	60	20	40	60	20	40	60
200	20		**0.00**	0.98	1.00	1.00	1.00	0.10	**0.00**	**0.00**	**0.03**	**0.00**	**0.00**	**0.03**	**0.00**	**0.00**	**0.00**	**0.00**	**0.00**
200	40	**0.00**		**0.00**	**0.00**	**0.00**	**0.00**	0.32	1.00	1.00	0.64	1.00	1.00	0.66	**0.00**	**0.00**	1.00	**0.02**	**0.00**
200	60	0.98	**0.00**		0.91	1.00	1.00	0.88	0.06	0.06	0.60	**0.00**	**0.00**	0.57	**0.00**	**0.00**	0.05	**0.00**	**0.00**
220	20	1.00	**0.00**	0.91		1.00	1.00	**0.04**	**0.00**	**0.00**	**0.01**	**0.00**	**0.00**	**0.01**	**0.00**	**0.00**	**0.00**	**0.00**	**0.00**
220	40	1.00	**0.00**	1.00	1.00		1.00	0.15	**0.00**	**0.00**	0.05	**0.00**	**0.00**	**0.05**	**0.00**	**0.00**	**0.00**	**0.00**	**0.00**
220	60	1.00	**0.00**	1.00	1.00	1.00		0.15	**0.00**	**0.00**	**0.05**	**0.00**	**0.00**	**0.04**	**0.00**	**0.00**	**0.00**	**0.00**	**0.00**
240	20	0.10	0.32	0.88	**0.04**	0.15	0.15		0.95	0.94	1.00	0.37	**0.02**	1.00	**0.00**	**0.00**	0.93	**0.00**	**0.00**
240	40	**0.00**	1.00	0.06	**0.00**	**0.00**	**0.00**	0.95		1.00	1.00	1.00	0.59	1.00	**0.00**	**0.00**	1.00	**0.00**	**0.00**
240	60	**0.00**	1.00	0.06	**0.00**	**0.00**	**0.00**	0.94	1.00		1.00	1.00	0.61	1.00	**0.00**	**0.00**	1.00	**0.00**	**0.00**
260	20	**0.03**	0.64	0.60	**0.01**	0.05	**0.05**	1.00	1.00	1.00		0.69	0.06	1.00	**0.00**	**0.00**	1.00	**0.00**	**0.00**
260	40	**0.00**	1.00	**0.00**	**0.00**	**0.00**	**0.00**	0.37	1.00	1.00	0.69		0.99	0.71	0.06	**0.00**	1.00	**0.02**	**0.00**
260	60	**0.00**	1.00	**0.00**	**0.00**	**0.00**	**0.00**	**0.02**	0.59	0.61	0.06	0.99		0.07	0.06	**0.00**	0.65	0.35	**0.00**
280	20	**0.03**	0.66	0.57	**0.01**	**0.05**	**0.04**	1.00	1.00	1.00	1.00	0.71	0.07		**0.00**	**0.00**	1.00	**0.00**	**0.00**
280	40	**0.00**	**0.00**	**0.00**	**0.00**	**0.00**	**0.00**	**0.00**	**0.00**	**0.00**	**0.00**	**0.00**	0.06	**0.00**		**0.01**	**0.00**	1.00	**0.02**
280	60	**0.00**	**0.00**	**0.00**	**0.00**	**0.00**	**0.00**	**0.00**	**0.00**	**0.00**	**0.00**	**0.00**	**0.00**	**0.00**	**0.01**		**0.00**	**0.00**	1.00
300	20	**0.00**	1.00	0.05	**0.00**	**0.00**	**0.00**	0.93	1.00	1.00	1.00	1.00	0.65	1.00	**0.00**	**0.00**		**0.00**	**0.00**
300	40	**0.00**	**0.02**	**0.00**	**0.00**	**0.00**	**0.00**	**0.00**	**0.00**	**0.00**	**0.00**	**0.02**	0.35	**0.00**	1.00	**0.00**	**0.00**		**0.00**
300	60	**0.00**	**0.00**	**0.00**	**0.00**	**0.00**	**0.00**	**0.00**	**0.00**	**0.00**	**0.00**	**0.00**	**0.00**	**0.00**	**0.02**	1.00	**0.00**	**0.00**	

Table A5. Analysis of variance for combustible parts (*CP*).

Tukey Test for *CP*; a Bold Font Signifies Statistically Significant Difference ($p < 0.05$)		200	200	200	220	220	220	240	240	240	260	260	260	280	280	280	300	300	300
		20	40	60	20	40	60	20	40	60	20	40	60	20	40	60	20	40	60
200	20		**0.00**	0.97	1.00	1.00	1.00	0.07	**0.00**	**0.00**	**0.00**	**0.00**	**0.00**	**0.02**	**0.00**	**0.00**	**0.00**	**0.00**	**0.00**
200	40	**0.00**		**0.00**	**0.00**	**0.00**	**0.00**	0.25	1.00	1.00	**0.00**	1.00	0.99	0.58	**0.00**	**0.00**	1.00	**0.01**	**0.00**
200	60	0.97	**0.00**		0.87	0.99	0.99	0.84	**0.04**	**0.04**	**0.00**	**0.00**	**0.00**	**0.49**	**0.00**	**0.00**	**0.03**	**0.00**	**0.00**
220	20	1.00	**0.00**	0.87		1.00	1.00	**0.03**	**0.00**	**0.00**	**0.00**	**0.00**	**0.00**	**0.01**	**0.00**	**0.00**	**0.00**	**0.00**	**0.00**
220	40	1.00	**0.00**	0.99	1.00		1.00	0.11	**0.00**	**0.00**	**0.00**	**0.00**	**0.00**	**0.03**	**0.00**	**0.00**	**0.00**	**0.00**	**0.00**
220	60	1.00	**0.00**	0.99	1.00	1.00		0.11	**0.00**	**0.00**	**0.00**	**0.00**	**0.00**	**0.03**	**0.00**	**0.00**	**0.00**	**0.00**	**0.00**
240	20	0.07	0.25	0.84	**0.03**	0.11	0.11		0.93	0.92	**0.00**	0.29	**0.01**	1.00	**0.00**	**0.00**	0.90	**0.00**	**0.00**
240	40	**0.00**	1.00	**0.04**	**0.00**	**0.00**	**0.00**	0.93		1.00	**0.00**	1.00	0.51	1.00	**0.00**	**0.00**	1.00	**0.00**	**0.00**
240	60	**0.00**	1.00	**0.04**	**0.00**	**0.00**	**0.00**	0.92	1.00		**0.00**	1.00	0.53	1.00	**0.00**	**0.00**	1.00	**0.00**	**0.00**
260	20	**0.00**	**0.00**	**0.00**	**0.00**	**0.00**	**0.00**	**0.00**	**0.00**	**0.00**		**0.00**	**0.00**	**0.00**	**0.00**	**0.00**	**0.00**	**0.00**	**0.00**
260	40	**0.00**	1.00	**0.00**	**0.00**	**0.00**	**0.00**	0.29	1.00	1.00	**0.00**		0.99	0.64	**0.00**	**0.00**	1.00	**0.01**	**0.00**
260	60	**0.00**	0.99	**0.00**	**0.00**	**0.00**	**0.00**	**0.01**	0.51	0.53	**0.00**	0.99		**0.04**	**0.04**	**0.00**	0.57	0.27	**0.00**
280	20	**0.02**	0.58	0.49	**0.01**	**0.03**	**0.03**	1.00	1.00	1.00	**0.00**	0.64	**0.04**		**0.00**	**0.00**	1.00	**0.00**	**0.00**
280	40	**0.00**	**0.00**	**0.00**	**0.00**	**0.00**	**0.00**	**0.00**	**0.00**	**0.00**	**0.00**	**0.00**	**0.04**	**0.00**		**0.01**	**0.00**	1.00	**0.01**
280	60	**0.00**	**0.00**	**0.00**	**0.00**	**0.00**	**0.00**	**0.00**	**0.00**	**0.00**	**0.00**	**0.00**	**0.00**	**0.00**	**0.01**		**0.00**	**0.00**	1.00
300	20	**0.00**	1.00	**0.03**	**0.00**	**0.00**	**0.00**	0.90	1.00	1.00	**0.00**	1.00	0.57	1.00	**0.00**	**0.00**		**0.00**	**0.00**
300	40	**0.00**	**0.01**	**0.00**	**0.00**	**0.00**	**0.00**	**0.00**	**0.00**	**0.00**	**0.00**	**0.01**	0.27	**0.00**	1.00	**0.00**	**0.00**		**0.00**
300	60	**0.00**	**0.00**	**0.00**	**0.00**	**0.00**	**0.00**	**0.00**	**0.00**	**0.00**	**0.00**	**0.00**	**0.00**	**0.00**	**0.01**	1.00	**0.00**	**0.00**	

Table A6. Analysis of variance for high heating value (HHV).

Tukey Test for HHV; a Bold Font Signifies Statistically Significant Difference ($p < 0.05$)		200	200	200	220	220	220	240	240	240	260	260	260	280	280	280	300	300	300
		20	40	60	20	40	60	20	40	60	20	40	60	20	40	60	20	40	60
200	20		0.05	1.00	1.00	1.00	1.00	0.99	0.31	**0.00**	0.87	0.76	0.10	0.99	**0.00**	**0.00**	0.97	**0.00**	**0.00**
200	40	0.05		**0.02**	0.13	0.18	**0.04**	0.62	1.00	1.00	0.93	0.97	1.00	0.62	0.78	**0.05**	0.77	0.98	**0.00**
200	60	1.00	**0.02**		1.00	1.00	1.00	0.95	0.15	**0.00**	0.66	0.52	**0.04**	0.95	**0.00**	**0.00**	0.87	**0.00**	**0.00**
220	20	1.00	0.13	1.00		1.00	1.00	1.00	0.55	**0.01**	0.98	0.94	0.22	1.00	**0.00**	**0.00**	1.00	**0.00**	**0.00**
220	40	1.00	0.18	1.00	1.00		1.00	1.00	0.68	**0.02**	0.99	0.98	0.31	1.00	**0.00**	**0.00**	1.00	**0.00**	**0.00**
220	60	1.00	**0.04**	1.00	1.00	1.00		0.99	0.25	**0.00**	0.82	0.69	0.07	0.99	**0.00**	**0.00**	0.95	**0.00**	**0.00**
240	20	0.99	0.62	0.95	1.00	1.00	0.99		0.98	0.14	1.00	1.00	0.80	1.00	**0.01**	**0.00**	1.00	**0.03**	**0.00**
240	40	0.31	1.00	0.15	0.55	0.68	0.25	0.98		0.94	1.00	1.00	1.00	0.98	0.25	**0.01**	1.00	0.59	**0.00**
240	60	**0.00**	1.00	**0.00**	**0.01**	**0.02**	**0.00**	0.14	0.94		0.41	0.54	1.00	0.14	1.00	0.33	0.22	1.00	**0.01**
260	20	0.87	0.93	0.66	0.98	0.99	0.82	1.00	1.00	0.41		1.00	0.98	1.00	**0.03**	**0.00**	1.00	0.12	**0.00**
260	40	0.76	0.97	0.52	0.94	0.98	0.69	1.00	1.00	0.54	1.00		1.00	1.00	0.05	**0.00**	1.00	0.18	**0.00**
260	60	0.10	1.00	**0.04**	0.22	0.31	0.07	0.80	1.00	1.00	0.98	1.00		0.79	0.60	**0.02**	0.90	0.91	**0.00**
280	20	0.99	0.62	0.95	1.00	1.00	0.99	1.00	0.98	0.14	1.00	1.00	0.79		**0.01**	**0.00**	1.00	**0.03**	**0.00**
280	40	**0.00**	0.78	**0.00**	**0.00**	**0.00**	**0.00**	**0.01**	0.25	1.00	**0.03**	0.05	0.60	**0.01**		0.97	**0.01**	1.00	0.20
280	60	**0.00**	**0.05**	**0.00**	**0.00**	**0.00**	**0.00**	**0.00**	**0.01**	0.33	**0.00**	**0.00**	**0.02**	**0.00**	0.97		**0.00**	0.75	0.98
300	20	0.97	0.77	0.87	1.00	1.00	0.95	1.00	1.00	0.22	1.00	1.00	0.90	1.00	**0.01**	**0.00**		0.05	**0.00**
300	40	**0.00**	0.98	**0.00**	**0.00**	**0.00**	**0.00**	**0.03**	0.59	1.00	0.12	0.18	0.91	**0.03**	1.00	0.75	0.05		0.06
300	60	**0.00**	**0.00**	**0.00**	**0.00**	**0.00**	**0.00**	**0.00**	**0.00**	**0.01**	**0.00**	**0.00**	**0.00**	**0.00**	0.20	0.98	**0.00**	0.06	

Table A7. Statistical evaluation of specific heat of elephant dung.

Intercept/ Coefficient	Value of Intercept/ Coefficient	Standard Error	p	Lower Limit of Confidence	Upper Limit of Confidence
a_1	7.74×10^0	2.64×10^{-1}	0.00	6.59×10^0	7.62×10^0
a_2	-6.55×10^{-1}	2.37×10^{-2}	0.00	-6.55×10^{-1}	-5.62×10^{-1}
a_3	2.37×10^{-2}	8.21×10^{-4}	0.00	2.09×10^{-2}	2.41×10^{-2}
a_4	-3.97×10^{-4}	1.45×10^{-5}	0.00	-4.11×10^{-4}	-3.54×10^{-4}
a_5	3.63×10^{-6}	0.00×10^0	0.00	3.53×10^{-6}	3.53×10^{-6}
a_6	-1.93×10^{-8}	0.00×10^0	0.00	-1.90×10^{-8}	-1.90×10^{-8}
a_7	6.04×10^{-11}	0.00×10^0	0.00	5.97×10^{-11}	5.97×10^{-11}
a_8	-1.03×10^{-13}	0.00×10^0	0.00	-1.02×10^{-13}	-1.02×10^{-13}
a_9	7.37×10^{-17}	0.00×10^0	0.00	7.37×10^{-17}	7.37×10^{-17}

$$SH = a_1 + a_2 \cdot T + a_3 \cdot T^2 + a_4 \cdot T^3 + a_5 \cdot T^4 + a_6 \cdot T^5 + a_7 \cdot T^6 + a_8 \cdot T^7 + a_9 \, T^8, \ R^2 = 0.98, \ R = 0.99.$$

Table A8. Evaluation of commercial pellet HHV_{daf}, based on [51].

Type of Pellet	Ash, %	HHV, MJ·kg^{-1}	HHV_{daf} *, MJ·kg^{-1}
Pine sawdust	0.66	19.52	19.65
Wheat straw	7.27	17.57	18.95
Corn settlements	1.27	18.80	19.04
Agricultural residues	8.27	18.13	19.76

* HHV_{daf} has been calculated based on Equation (4).

References

1. Brown, J.L.; Paris, S.; Prado-Oviedo, N.A.; Meehan, C.L.; Hogan, J.N.; Morfeld, K.A.; Carlstead, K. Reproductive Health Assessment of Female Elephants in North American Zoos and Association of Husbandry Practices with Reproductive Dysfunction in African Elephants (Loxodonta africana). *PLoS ONE* **2016**, *11*, e0145673. [CrossRef] [PubMed]

2. Farah, N.; Amna, M.; Naila, Y.; Ishtiaq, R. Processing of Elephant Dung and its Utilization as a Raw Material for Making Exotic Paper. *Res. J. Chem. Sci.* **2014**, *4*, 94–103.

3. Sannigrahi, A.K. Beneficial Utilization Of Elephant Dung Through Vermicomposting. *Int. J. Recent Sci. Res.* **2015**, *6*, 4814–4817.

4. Elephants for Africa. Available online: http://www.elephantsforafrica.org/elephant-facts/ (accessed on 18 August 2019).

5. Schröder, J.J.; Scholefield, D.; Cabral, F.; Hofman, G. The effects of nutrient losses from agriculture on ground and surface water quality: The position of science in developing indicators for regulation. *Environ. Sci. Policy* **2004**, *7*, 15–23. [CrossRef]

6. Pérez-Godínez, E.A.; Lagunes-Zarate, J.; Corona-Hernández, J.; Barajas-Aceves, M. Growth and reproductive potential of Eisenia foetida (Sav) on various zoo animal dungs after two methods of pre-composting followed by vermicomposting. *Waste Manag.* **2017**, *64*, 67–78. [CrossRef]

7. Zhou, S.; Liang, H.; Han, L.; Huang, G.; Yang, Z. The influence of manure feedstock, slow pyrolysis, and hydrothermal temperature on manure thermochemical and combustion properties. *Waste Manag.* **2019**, *88*, 85–95. [CrossRef]

8. Qambrani, N.A.; Rahman, M.M.; Won, S.; Shim, S.; Ra, C. Biochar properties and eco-friendly applications for climate change mitigation, waste management, and wastewater treatment: A review. *Renew. Sustain. Energy Rev.* **2017**, *79*, 255–273. [CrossRef]

9. Aira, M.; Monroy, F.; Domínguez, J.; Mato, S. How earthworm density affects microbial biomas and activity in pig manure. *Eur. J. Soil Biol.* **2002**, *38*, 7–10. [CrossRef]

10. Sukasem, N.; Khanthi, K.; Prayoonkham, S. Biomethane Recovery from Fresh and Dry Water Hyacinth Anaerobic Co-Digestion with Pig Dung, Elephant Dung and Bat Dung with Different Alkali Pretreatments. *Energy Procedia* **2017**, *138*, 294–300. [CrossRef]

11. Klasson, K.T.; Nghiem, N. *Energy Production from Zoo Animal Wastes*; Oak Ridge National Laboratory: Oak Ridge, TN, USA, 2003; Volume 3, pp. 1–8. [CrossRef]

12. Fangkum, A.; Reungsang, A. Simultaneous saccharification and fermentation of cellulose for bio-hydrogen production by anaerobic mixed cultures in elephant dung. *Int. J. Hydrogen Energy* **2014**, *39*, 9028–9035. [CrossRef]

13. Fangkum, A.; Reungsang, A. Biohydrogen production from mixed xylose/arabinose at thermophilic temperature by anaerobic mixed cultures in elephant dung. *Int. J. Hydrogen Energy* **2011**, *36*, 13928–13938. [CrossRef]

14. Moosophin, K.; Phachan, N.; Apiraksakorn, J. Screening of cellulolytic clostridia from animal dung and compost for direct butanol production from cellulosic materials. *Curr. Opin. Biotechnol.* **2013**, *24*, 48–143. [CrossRef]

15. Li, J.; Xiao, F.; Zhang, L.; Amirkhanian, S.N. Life cycle assessment and life cycle cost analysis of recycled solid waste materials in highway pavement: A review. *J. Clean. Prod.* **2019**, *233*, 1182–1206. [CrossRef]

16. Jia, J.; Shu, L.; Zang, G.; Xu, L.; Abudula, A.; Ge, K. Energy analysis of a co-gasification of woody biomass and animal manure, solid oxide fuel cells and micro gas turbine hybrid system. *Energy* **2018**, *149*, 750–761. [CrossRef]

17. Das, K.; Hiloidhari, M.; Baruah, D.C.; Nonhebel, S. Impact of Time Expenditure on Household Preferences for Cooking Fuels. *Energy* **2018**, *151*, 309–316. [CrossRef]

18. International Monetary Fund Report for Selected Countries and Subjects. Available online: https://data.worldbank.org/indicator/SP.POP.TOTL?name_desc=false (accessed on 23 August 2019).

19. Ji, C.; Cheng, K.; Nayak, D.; Pan, G. Environmental and economic assessment of crop residue competitive utilization for biochar, briquette fuel and combined heat and power generation. *J. Clean. Prod.* **2018**, *192*, 916–923. [CrossRef]

20. Kamara, J.; Galukande, M.; Maeda, F.; Luboga, S.; Renzaho, A. Understanding the Challenges of Improving Sanitation and Hygiene Outcomes in a Community Based Intervention: A Cross-Sectional Study in Rural Tanzania. *Int. J. Environ. Res. Public Health* **2017**, *14*, 602. [CrossRef]

21. Białowiec, A.; Micuda, M.; Szumny, A.; Łyczko, J.; Koziel, J.A. Waste to Carbon: Influence of Structural Modification on VOC Emission Kinetics from Stored Carbonized Refuse-Derived Fuel. *Sustainability* **2019**, *11*, 935. [CrossRef]

22. Stępień, P.; Białowiec, A. Kinetic Parameters of Torrefaction Process of Alternative Fuel Produced From Municipal Solid Waste and Characteristic of Carbonized Refuse Derived Fuel. *Detritus* **2018**, *3*, 75–83. [CrossRef]

23. Waste Characteristics. Calculation of Dry Mass on the Basis of Dry Residue or Water Content. Polish standard PN-EN 14346:2011. Available online: https://infostore.saiglobal.com/enau/Standards/PN-EN-14346-2011-932471_SAIG_PKN_PKN_2197939/ (accessed on 23 August 2019).

24. Waste Characteristics. Content of Organic Matter. Polish Standard PN-EN 15169:2011. Available online: http://sklep.pkn.pl/pn-en-15169-2011p.html (accessed on 23 August 2019).

25. Municipal Solid Waste. Combustible and Non-Combustible Content. Polish Standard PN-Z-15008-04:1993. Available online: http://sklep.pkn.pl/pn-z-15008-04-1993p.html (accessed on 23 August 2019).

26. Solid Fuels. Determination of the Higher Heating Value and the Lower Heating Value. Polish Standard PN-G-04513:1981. Available online: http://sklep.pkn.pl/pn-g-04513-1981p.html (accessed on 23 August 2019).

27. Chin, K.L.; H'ng, P.S.; Go, W.Z.; Wong, W.Z.; Lim, T.W.; Maminski, M.; Paridah, M.T.; Luqman, A.C. Optimization of torrefaction conditions for high energy density solid biofuel from oil palm biomass and fast growing species available in Malaysia. *Ind. Crops Prod.* **2013**, *49*, 768–774. [CrossRef]

28. Pulka, J.; Wiśniewski, D.; Gołaszewski, J.; Białowiec, A. Is the biochar produced from sewage sludge a good quality solid fuel? *Arch. Environ. Prot.* **2016**, *42*, 125–134. [CrossRef]

29. Peleg, M.; Normand, M.D.; Corradini, M.G. The Arrhenius Equation Revisited. *Crit. Rev. Food Sci. Nutr.* **2012**, *52*, 830–851. [CrossRef] [PubMed]

30. Stępień, P.; Serowik, M.; Koziel, J.A.; Białowiec, A. Waste to carbon energy demand model and data based on the TGA and DSC analysis of individual MSW components. *Data* **2019**, *4*, 53. [CrossRef]

31. Stępień, P.; Mysior, M.; Białowiec, A. Technical and technological problems and potential application waste torrefaction. In *Innovations in Waste Management—Selected Issues*; Manczarski, P., Ed.; Wydawnictwo Uniwersytetu Przyrodniczego: Wrocław, Poland, 2018; Volume 1, pp. 59–78. Available online: https://www.researchgate.net/publication/325367684_Innowacje_w_gospodarce_odpadami_Zagadnienia_wybrane (accessed on 23 August 2019).

32. Świechowski, K.; Liszewski, M.; Bąbelewski, P.; Koziel, J.A.; Białowiec, A. Oxytree Pruned Biomass Torrefaction: Mathematical Models of the Influence of Temperature and Residence Time on Fuel Properties Improvement. *Materials* **2019**, *12*, 2228. [CrossRef] [PubMed]

33. Li, S.; Harris, S.; Anandhi, A.; Chen, G. Predicting biochar properties and functions based on feedstock and pyrolysis temperature: A review and data syntheses. *J. Clean. Prod.* **2019**, *215*, 890–902. [CrossRef]

34. Kim, D.; Lee, K.; Park, K.Y. Upgrading the characteristics of biochar from cellulose, lignin, and xylan for solid biofuel production from biomass by hydrothermal carbonization. *J. Ind. Eng. Chem.* **2016**, *42*, 95–100. [CrossRef]

35. Gascó, G.; Paz-Ferreiro, J.; Álvarez, M.L.; Saa, A.; Méndez, A. Biochars and hydrochars prepared by pyrolysis and hydrothermal carbonisation of pig manure. *Waste Manag.* **2018**, *79*, 395–403. [CrossRef]

36. Van der Stelt, M.J.C. *Chemistry and Reaction Kinetics of Biowaste Torrezfaction*; Technische Universiteit Eindhoven: Eindhoven, The Netherlands, 2011. [CrossRef]

37. Pahla, G.; Mamvura, T.A.; Ntuli, F.; Muzenda, E. Energy densification of animal waste lignocellulose biomass and raw biomass. *S. Afr. J. Chem. Eng.* **2017**, *24*, 168–175. [CrossRef]

38. Higgins, M.J.; Adams, G.; Chen, Y.-C.; Erdal, Z.; Forbes, R.H.; Glindemann, D.; Hargreaves, J.R.; McEwen, D.; Murthy, S.N.; Novak, J.T.; et al. Role of protein, amino acids, and enzyme activity on odor production from anaerobically digested and dewatered biosolids. *Water Environ. Res.* **2008**, *80*, 127–135. [CrossRef]

39. Lehmann, J.; Joseph, S. *Biochar for Environmental Management: Science and Technology*; Earthscan: London, UK, 2009; ISBN 184407658X.

40. Ro, K.S.; Libra, J.A.; Bae, S.; Berge, N.D.; Flora, J.R.V.; Pecenka, R. Combustion Behavior of Animal-Manure-Based Hydrochar and Pyrochar. *ACS Sustain. Chem. Eng.* **2019**, *7*, 470–478. [CrossRef]

41. Avellone, E.A.; Baumeister, T.; Saunders, H. *Marks Standard Handbook for Mechanical Engineers*; McGraw-Hill Education: New York, NY, USA, 2008; pp. 101–105.

42. Oshita, K.; Toda, S.; Takaoka, M.; Kanda, H.; Fujimori, T.; Matsukawa, K.; Fujiwara, T. Solid fuel production from cattle manure by dewatering using liquefied dimethyl ether. *Fuel* **2015**, *159*, 7–14. [CrossRef]

43. Dębska, A.; Koziołek, S.; Bieniek, J.; Białowiec, A. Potencjał produkcji biogazu z odpadów we wrocławskim Ogrodzie Zoologicznym. *Annu. Set Environ. Prot.* **2016**, *18*, 337–351.

44. Matthiessen, M.K.; Larney, F.J.; Selinger, L.B.; Olson, A.F. Influence of Loss-on-Ignition Temperature and Heating Time on Ash Content of Compost and Manure. *Commun. Soil Sci. Plant Anal.* **2005**, *36*, 2561–2573. [CrossRef]

45. Świechowski, K.; Koziel, J.A.; Liszewski, M.; Bąbelewski, P.; Białowiec, A. Fuel Properties of Torrefied Biomass from Pruning of Oxytree. *Data* **2019**, *4*, 55. [CrossRef]

46. Dudek, M.; Świechowski, K.; Manczarski, P.; Koziel, J.A.; Białowiec, A. The Effect of Biochar Addition on the Biogas Production Kinetics from the Anaerobic Digestion of Brewers' Spent Grain. *Energies* **2019**, *12*, 1518. [CrossRef]

47. Gautam, S.; Edwards, R.; Yadav, A.; Weltman, R.; Pillarsetti, A.; Arora, N.K.; Smith, K.R. Probe-based measurements of moisture in dung fuel for emissions measurements. *Energy Sustain. Dev.* **2016**, *35*, 1–6. [CrossRef]

48. Lang, Q.; Guo, Y.; Zheng, Q.; Liu, Z.; Gai, C. Co-hydrothermal carbonization of lignocellulosic biomass and swine manure: Hydrochar properties and heavy metal transformation behavior. *Bioresour. Technol.* **2018**, *266*, 242–248. [CrossRef]

49. Rousset, P.; Aguiar, C.; Labbé, N.; Commandré, J.-M. Enhancing the combustible properties of bamboo by torrefaction. *Bioresour. Technol.* **2011**, *102*, 8225–8231. [CrossRef]

50. Bach, Q.-V.; Skreiberg, Ø. Upgrading biomass fuels via wet torrefaction: A review and comparison with dry torrefaction. *Renew. Sustain. Energy Rev.* **2016**, *54*, 665–677. [CrossRef]

51. Dyjakon, A.; Noszczyk, T. The Influence of Freezing Temperature Storage on the Mechanical Durability of Commercial Pellets from Biomass. *Energies* **2019**, *12*, 2627. [CrossRef]

52. Lacey, J.A.; Aston, J.E.; Westover, T.L.; Cherry, R.S.; Thompson, D.N. Removal of introduced inorganic content from chipped forest residues via air classification. *Fuel* **2015**, *160*, 265–273. [CrossRef]

53. Várhegyi, G.; Bobály, B.; Jakab, E.; Chen, H. Thermogravimetric study of biomass pyrolysis kinetics. A distributed activation energy model with prediction tests. *Energy Fuels* **2011**, *25*, 24–32. [CrossRef]

54. Bach, V.; Tran, K.-Q. Dry and wet torrefaction of woody biomass—A comparative study on combustion kinetics. *Energy Proceedia* **2015**, *75*, 150–155. [CrossRef]

55. Soria-Verdugo, A.; Goos, E.; García-Hernando, N. Effect of the number of TGA curves employed on the biomass pyrolysis kinetics results obtained using the Distributed Activation Energy Model. *Fuel Process. Technol.* **2015**, *134*, 360–371. [CrossRef]
56. Pulka, J.; Manczarski, P.; Koziel, J.A.; Białowiec, A. Torrefaction of Sewage Sludge: Kinetics and Fuel Properties of Biochars. *Energies* **2019**, *12*, 565. [CrossRef]
57. Syguła, E.; Koziel, J.A.; Białowiec, A. Proof-of-Concept of Spent Mushrooms Compost Torrefaction—Studying the Process Kinetics and the Influence of Temperature and Duration on the Calorific Value of the Produced Biocoal. *Energies* **2019**, *12*, 3060. [CrossRef]
58. Basak, S.; Samanta, K.K. Thermal behaviour and the cone calorimetric analysis of the jute fabric treated in different pH condition. *J. Therm. Anal. Calorim.* **2019**, *135*, 3095–3105. [CrossRef]
59. Wang, J.; Wang, G.; Zhang, M.; Chen, M.; Li, D.; Min, F.; Chen, M.; Zhang, S.; Ren, Z.; Yan, Y. A comparative study of thermolysis characteristics and kinetics of seaweeds and fir wood. *Process Biochem.* **2006**, *41*, 1883–1886. [CrossRef]
60. Said, N.; Abdel daiem, M.M.; García-Maraver, A.; Zamorano, M. Reduction of Ash Sintering Precursor Components in Rice Straw by Water Washing. *BioResources* **2014**, *9*, 6756–6764. [CrossRef]
61. Burhenne, L.; Messmer, J.; Aicher, T.; Laborie, M.-P. The effect of the biomass components lignin, cellulose and hemicellulose on TGA and fixed bed pyrolysis. *J. Anal. Appl. Pyrolysis* **2013**, *101*, 177–184. [CrossRef]

Article

Generalized Energy and Ecological Characteristics of the Process of Co-Firing Coal with Biomass in a Steam Boiler

Joachim Kozioł [1], Joanna Czubala [2], Michał Kozioł [3] and Piotr Ziembicki [1,*]

[1]   Faculty of Civil Engineering, Architecture and Environmental Engineering, University of Zielona Góra, Prof. Z. Szafrana 15 St., 65-516 Zielona Góra, Poland; kojo643@interia.pl

[2]   TAURON Polish Energy Inc., Ściegiennego 3 St., 40-114 Katowice, Poland; joanna.czubala@wp.pl

[3]   Faculty of Energy and Environmental Engineering, Silesian University of Technology, 44-100 Gliwice, Poland; michal.koziol@polsl.pl

*   Correspondence: p.ziembicki@iis.uz.zgora.pl

Received: 18 April 2020; Accepted: 15 May 2020; Published: 21 May 2020

Abstract: One of the ways used to reduce the emission of carbon dioxide and other harmful substances is the implementation of biomass co-firing processes with coals. Such processes have been implemented for many years throughout many countries of the world, and have included using existing high-power coal boilers. Despite numerous experiments, there are still no analyses in the literature allowing for their generalization. The purpose of this paper is to determine the generalized energy and ecological characteristics of dust steam boilers co-firing hard coal with biomass. The energy characteristics determined in the paper are the dependence of the gross energy efficiency of boilers on such decision parameters as their efficiency and the share of biomass chemical energy in fuel. However, the ecological characteristics are the dependence of emission streams: CO, NO_x, SO_2, and dust on the same decision parameters. From a mathematical point of view, the characteristics are approximation functions between the efficiency values obtained from the measurements and the emission streams of the analysed harmful substances and the corresponding values of the decision parameters. Second-degree polynomials are assumed in this paper as approximation functions. Therefore, determining the characteristics came down to determining the constant coefficients occurring in these polynomials, the so-called structural parameters. The fit of the determined characteristics was assessed based on the coefficients of random variation and the test of estimated significance of structural parameters. Boiler characteristics can be used when forecasting the impact of changes in operating conditions on the effects achieved in existing, modernized, and designed boilers. The generalization of the characteristics was obtained from the measurement results presented in 10 independent sources used to determine them.

Keywords: steam boilers; co-firing; biomass; characteristics; boiler efficiency; GHG emissions; decision parameters; result parameters; structural parameters

---

1. Introduction

The specificity of the Polish energy system is one of the highest in the world's share of sources using coal (both hard and brown) in electricity production. In 2018, this share amounted to 78.2% [1]. For comparison, 12.8% of electricity came from renewable energy sources, wherein 3.9% came from sources using biomass (biogas plants, biomass combustion plants, and biomass co-firing) [1]. At the same time, the share of energy consumption from renewable sources in the entire energy balance in 2018 was 11.3% [2]. In 2017, solid biofuels accounted for 67.9% of energy obtained from renewable sources in Poland, and liquid biofuels accounted for 10% [3]. Due to the geographical location (relatively low sun exposure and only locally occurring areas with more frequent strong winds) and the country's surface (flat with a lack of larger numbers of sizable rivers), until recently, hopes for the development of renewable energy sources in Poland were mainly associated with sources using biomass. Currently, due to the development of technologies in the field of solar energy, and in connection with projects to build a number of wind installations in the Baltic Sea, the prospective importance of these sources has grown.

Primary energy obtained in Poland from biomass sources corresponds to approximately 7.5 Mtoe [3]. This is a small amount in relation to the national biomass energy potential. This potential results from developed agriculture, using approximately 188 thousand km^2 of arable land (nearly 60% of the country), and forest areas of 95 thousand km^2 (30% of the country area) [4]. The technical potential of biomass in Poland is estimated at around 930 PJ/year [5]. This corresponds to almost 22.2 Mtoe per year. This potential mainly consists of waste wood from forests (200–240 PJ/year), energy crops (130 PJ/year), undeveloped straw (over 110 PJ/year), post-use wood (over 40 PJ/year), wastes from the wood industry (about 30 PJ/year), and biomass from agricultural sources (about 15 PJ/year) [5]. Thus, around 1/3 of the domestic biomass energy potential is currently used.

Biomass co-firing with coal in large power boilers developed dynamically in Poland over the years 2005 to 2012. In the peak period, the co-firing process was responsible for about half of the electricity generated from renewable sources in Poland. Economic considerations, based on the introduced legal regulations, spoke in favour of the process. After 2012, due to the suspension of government support for this process, co-firing in large energy facilities began decreasing, and now only about 10% of electricity from 2012 is generated in this process. However, currently due to difficulties in meeting the national target of the share of renewable energy sources in gross final consumption in 2020, under the objectives of the European Union (20 % EU, 15% Poland), and a new, higher target for 2030 (32% for the entire EU, no national target yet), the Polish government plans to allocate 10 billion PLN (over 2.2 billion EUR) to re-support this process [6].

It should be noted that although such a strong dependence on coal in the electricity sector is a Polish specificity, coal is still an important element of the power systems in a number of other countries, including Germany, China, Russia, India, Turkey, and South Africa. In addition, coal mining has increased in recent years, according to the International Energy Agency [7]. The organization forecasts that until 2024 its production will remain at a level similar to the current one. In addition, in non-OECD countries in 2017, electricity production from coal-fired installations accounted for 47% of the electricity produced [8].

Co-firing is a simple, cheap, and fast way to increase the production of electricity from renewable sources, as well as to reduce CO_2 emissions from existing coal-fired installations. Appropriate installations already exist, and as a result of many years of process implementation, relevant experience in the process has been acquired in a number of countries. Experiments related to biomass co-firing in power plants and combined heat and power plants have been extensively discussed, e.g., in review works [9–12]. The advantages of the process include most often the reduction of greenhouse gas emissions [9,10,13], and in the case of some installations and biomass, the reduction of NO_x emissions [9,10,14], as well as the economic efficiency of the process (most often resulting from administrative and legal support activities) [9,15,16].

Dust furnaces are one of the most frequently used constructions of large boilers for the implementation of co-firing of coals and biomass [9]. However, biomass co-firing in these units also has a number of disadvantages. These disadvantages concern both the installation itself and the entire biomass management. The latter include:

- Difficulties with the delivery of large amounts of biomass to individual co-firing installations (the need to transport biomass over long distances, including even its import);
- Environmental costs of biomass transport over considerable distances;

- Lack of raw material for other biomass recipients, because the financial possibilities and obtained margin on sold energy from the co-firing process often allows operators of these installations to purchase biomass at much higher prices than the competition;
- Competition between energy crops and food crops.

Regarding the process disadvantages occurring in the installation itself, these should be mentioned:

- Reduction of boiler efficiency [17,18];
- Deposit formation, corrosion, and erosion of boiler components [10,12,19];
- Problems with milling and operation of coal mills, and in the fuel feeding system (increased fire risk and explosion) [20,21].

There are various ways to evaluate the effects of the co-firing process. Among them we can distinguish among others:

- Assessment of economic effects [15–17,22];
- Application of LCA (Life Cycle Assessment) methodology [23,24];
- Assessment of technical and ecological aspects [10,12,18,19,21,22,25,26].

Data for conducted analyses are the result of process modeling, e.g., [14,22,27] or are obtained based on, e.g., measurements of emissions and installation efficiency during tests at industrial facilities [19,21,25]. As mentioned, many countries around the world are already very experienced in the implementation of co-firing coal and biomass in high power steam boilers. Despite this, however, there is a lack of literature attempting to generalize and mathematically describe it.

The main purpose of the study is to determine the generalized energy and ecological characteristics of co-firing coal with biomass in boilers with a capacity higher than 130 t/h of steam. The energy characteristics determine the impact of boiler efficiency and the biomass chemical energy stream (so-called decision parameters) on the gross energy efficiency of the boiler (so-called result parameter). On the other hand, ecological characteristics make it possible to assess the impact of these decision parameters on the corresponding resulting emissions of the following substances: carbon oxides (CO), nitrogen oxides (NO_x), sulphur dioxide (SO_2), and dust.

Energy characteristics can be used to control the operation of steam boilers and are necessary when determining the optimal operating conditions for boilers being designed and modernized. The generalization of the characteristics made in this paper consist of using the authors' own measurements and information given in the literature to determine them.

2. Materials and Methods

2.1. Method for Determining the Characteristics

The characteristics are mathematical dependencies (functions) of the result parameters on the decision parameters. In the analysed case, the considered functions can be written as:

$$\left.\begin{array}{l} y_1 \\ y_2 \\ y_3 \\ y_4 \\ y_5 \end{array}\right\} = f\left(x_1, x_2\right) \tag{1}$$

where: decision parameters: x_1—boiler efficiency, t/h, x_2—chemical biomass energy stream, MW; result parameters: y_1—boiler energy efficiency, y_2—emission of CO, kg/h, y_3—emission of NO_x, kg/h, y_4—emission of SO_2, kg/h, y_5—emission of dust, kg/h. The analysed problem is shown in Figure 1.

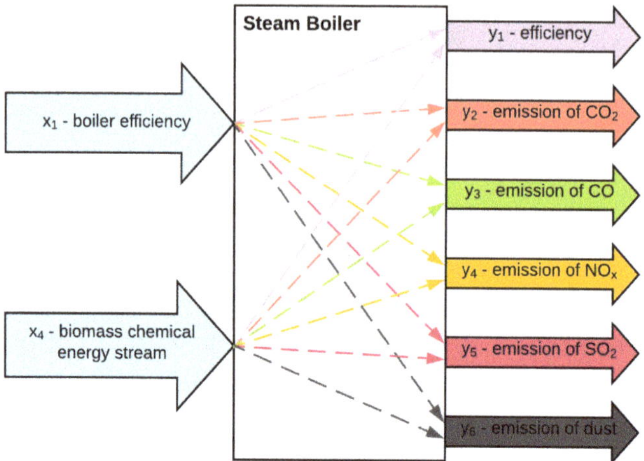

Figure 1. Characteristics of the test object.

The figure illustrates functional relationships between parameters resulting from decisive parameters. The characteristics are determined separately for individual analysed result parameters. These functions are approximation dependencies between the result parameter values obtained from the experiment and the combinations of decision parameter values characterizing them. Each of the combinations of decision values and the corresponding values of the resulting parameters is a measuring point.

It is recommended [28] to assume the form of object functions as second-degree polynomials. The following arguments support the use of a second order polynomial as a characteristic equation [28]:

- Its use is a compromise which, on the one hand, takes non-linearity of real test objects into account, and on the other hand, incorporates the tendency to reduce the number of measurements (an increase in the degree of polynomial significantly increases the number of tests necessary to carry out);
- It is characterized by high universality, as it allows adequate approximation of both simple linear and non-linear relationships;
- A polynomial is a particularly convenient form of a mathematical model in a case when only types of test factors and trends of their influence on the resulting factor are known.

In this case, it is proposed to take the following form of this function:

$$y_j = b_0^{(j)} + b_1^{(j)} x_1 + b_2^{(j)} x_2 + b_{11}^{(j)} x_1^2 + b_{22}^{(j)} x_2^2 + b_{12}^{(j)} x_1 x_2 \tag{2}$$

where: $b_0^{(j)}$, $b_1^{(j)}$, $b_2^{(j)}$, $b_{11}^{(j)}$, $b_{22}^{(j)}$, $b_{12}^{(j)}$—constant polynomial coefficients for the j-th result parameter, j—result parameter number ($j = 1 \div 5$).

According to the nomenclature used in statistical analyses in the remainder of the study, constant coefficients will be called structural parameters. Determining the characteristics comes down to determining the structural parameters in the individual characteristics. To determine the characteristics (function of the object), it is necessary to have a number of measurements that meet the condition:

$$N \geq N_b \tag{3}$$

where: N—number of measurements, N_b—number of structural parameters.

In the case under consideration, the number of measurements should be greater (or at least equal) to six. At the same time, the greater the number of measurements above the number of structural parameters, the greater the reliability of the determined coefficients. A larger number of measurements make it possible to reduce the uncertainty of measurements on the results of the calculations. The values of constant coefficients are determined by statistical methods. Most often, modified methods of least squares of deviations are used for this purpose [29–31]. Examples of determining the characteristics are given in [32,33].

2.2. Test Objects and Co-Firing Conditions

The research presented in the study concerns dust steam boilers with an efficiency of 130 to 230 t/h of superheated steam at a pressure of 10 or 13.6 MPa and a temperature of 450 or 540 °C, respectively. The amount of biomass chemical energy in co-fired fuel ranged from 0 to 100 MW, which, depending on the boiler efficiency, ranged from 0% to even 100% of the total chemical energy of the fuel.

In determining the characteristics, the results of experiments carried out by the authors [32] or described in the literature [19,27,34–40] were used. In order to determine the appropriate number of measurements and obtain the most representative results possible, the combinations of decision parameter values adopted for statistical analyses should be determined in accordance with the principles of scientific experiment planning [28–30,41]. Scientific planning of an experiment is possible when the determination of variants of systems of decision parameters (measuring points) precedes the measurements themselves. They are then carried out "on dictation" in strictly defined points of the research space (based on mathematical analyses). The authors attempted to adhere to these principles when determining the characteristics in the papers [32,33]. An analogous procedure was not possible when using the results of the experiments presented in the literature. In such cases, the values of decision parameters are determined by the authors "a priori" without scientific justification. At the same time, the use of information given in 10 independent sources to determine the characteristics allowed for generalization of the obtained results.

3. Results

3.1. Obtained Characteristics

Using the results of the experiments, structural parameters were estimated for each of the analysed object functions of the study. Table 1 shows the results obtained.

Table 1. Structural parameters of the characteristics.

No.	Coefficient Designation	Coefficient Values for				
		Boiler Energy Efficiency, $j = 1$	Emission of CO kg/h, $j = 2$	Emission of NO$_x$ kg/h, $j = 3$	Emission of SO$_2$ kg/h, $j = 4$	Emission of Dust kg/h, $j = 5$
1	b_0	0.8688	−311.8	1563	5320	−0.5138
2	b_1	0.0006	4.598	−13.47	−49.74	0.1185
3	b_2	−0.0006	−1.651	1.594	26.60	0.4437
4	b_{11}	-1.085×10^{-6}	−0.01261	0.03571	0.1300	−0.00011
5	b_{22}	4.236×10^{-6}	0.01471	−0.02809	−0.3471	−0.00317
6	b_{12}	-8.760×10^{-8}	0.001065	0.01429	0.08572	0.00082

3.2. Verification of Characteristics

The designated functions of the object were subjected to verification, which aimed to check whether the estimated functions describe the tested relationships well. The verification consisted of checking the following properties [42]:

1. The degree of compliance of the constructed model with the measured data using [42,43]:

   - Graphic comparison of the calculated values of the model with the measured data;
   - Residual variance;
   - Coefficient of random variation.

2. Quality assessment of structural parameters, requiring:

   - Determination of standard errors in the estimation of structural parameters;
   - Conducting the significance test of estimated structural parameters.

3.2.1. Verification of the Degree of Compliance of the Model with the Measurement Data

Graphic comparison of predicted values obtained from the characteristics with measured values is presented in Figures 2–6.

From the information given in Figure 2, there is a relatively high agreement between measured and predicted energy efficiency values. The value of an average standard deviation for individual characteristics is given in Table 2.

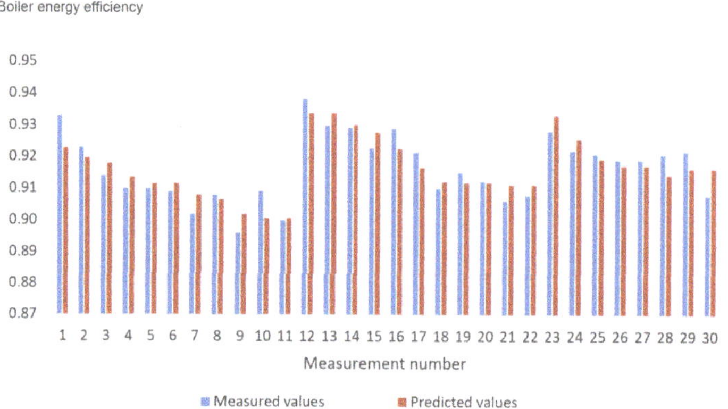

Figure 2. Comparison of the predicted values of the boiler energy efficiency model with measured values.

Table 2. Residual standard deviation.

	Values Corresponding to the Designation of				
	Boiler Energy Efficiency, $j = 1$	Emission of CO $j = 2$	Emission of NO_x $j = 3$	Emission of SO_2 $j = 4$	Emission of Dust $j = 5$
se^2	0.0000	1155	9084	529,000	71.07
se	0.01	33.99	95.31	727.3	8.43

Figures 3 and 4 show that the differences between the measured and predicted values of CO and NO_x emission streams are relatively large. This applies in particular to measurements 14 and 15. It should be presumed that the measuring points relate to objects with various operating conditions, including the ratio of excess air for combustion and the design of the burners.

Figures 5 and 6 show large differences between the measured and predicted SO_2 and dust emission streams. This applies, for example, to measurements 1, 12–15, 18, 20 and 24. It should be assumed that in these cases, decisive parameters should be supplemented with sulfur content and ash in coal and biomass. The authors were not in possession of such data.

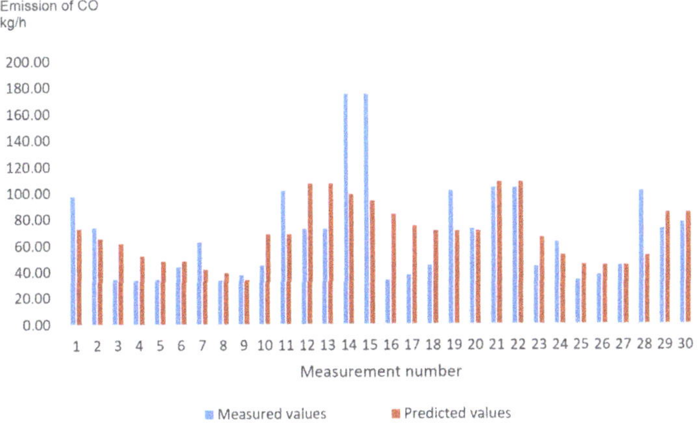

Figure 3. Comparison of the predicted values of the CO emission model with measured values.

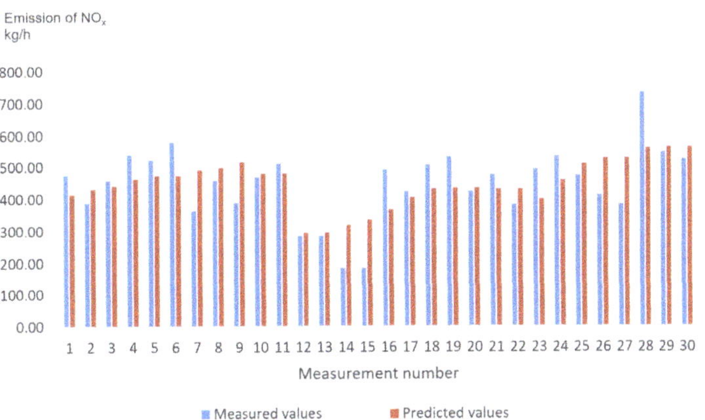

Figure 4. Comparison of the predicted values of the NO_x emission model with measured values.

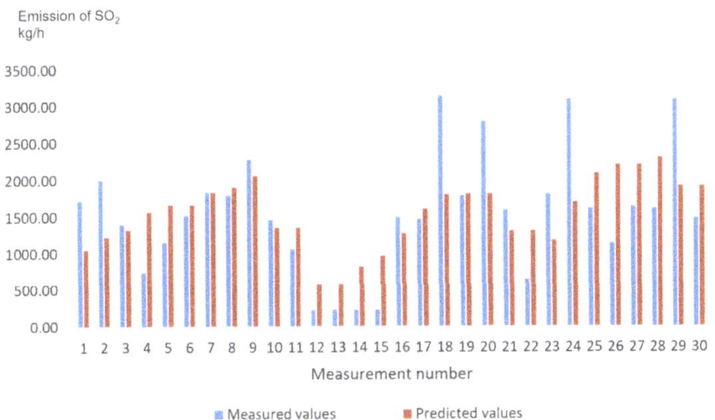

Figure 5. Comparison of the predicted values of the SO_2 emission model with measured values.

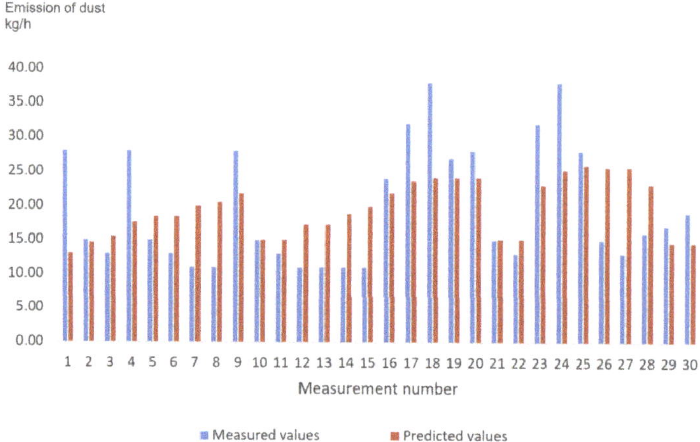

Figure 6. Comparison of the predicted values of the dust emission model with measured values.

Based on the graphical analysis of data obtained after estimating the structural parameters of the function, it can be seen that, in the case of boiler energy efficiency, they are well matched; in the case of ecological characteristics, the function matching is smaller.

Important information about the assessment of the compliance of the model with measured values is provided by the standard deviation of s residues and the coefficients of random variation v. These values were determined in accordance with the recommendations given in the literature [42,43]. The obtained results are given in Tables 2 and 3.

The values given in Table 2 were used for calculations of data presented in Table 3.

Table 3. Arithmetic mean values and random coefficient of variation.

	Values Corresponding to the Designation of				
	Boiler Energy Efficiency, $j = 1$	Emission of CO $j = 2$	Emission of NO$_x$ $j = 3$	Emission of SO$_2$ $j = 4$	Emission of Dust $j = 5$
$\bar{y}$	0.92	69.17	447.9	1548	19.63
v_e	0.01	0.49	0.21	0.47	0.43

The coefficient of random variation v is a relative measure of the diversity of the resulting parameters. It is the ratio of the standard deviation of the residuals and the arithmetic mean of the resulting parameter. It is assumed that its value should be less than 0.2. This condition is met by efficiency characteristics and possibly NO$_x$ emission characteristics

3.2.2. The Importance of Structural Parameters

According to information provided in the literature [28,29,31,42,44], structural parameter b is significant when the quotient of its value and the value of the average error of estimation of this parameter $s(b)$ (so-called Student test) is greater than the critical value t_c. The critical value t_c depends on the confidence level α and the number of degrees of freedom $m = N - k - 1$ (where: k—number of decision parameters). Its value can be read from commonly available tables. For $\alpha = 0.05$, $m = 30 - 2 - 1 = 27$, critical value $t_c = 2.052$.

Table 4 gives the values of the average error in estimating structural parameters $s(b)^{(j)}$. In turn, Table 5 gives the values of $t(b)^{(j)}$ for individual parameter.

Table 4. Standard errors in estimating the structural parameters of the test object functions.

No.	Indication of Estimate Error	Values of Errors in Determining				
		Boiler Energy Efficiency, $j = 1$	Emission of CO $j = 2$	Emission of NO_x $j = 3$	Emission of SO_2 $j = 4$	Emission of Dust $j = 5$
1	$s(b_0)$	0.0250	161.5	453.0	3457	40.06
2	$s(b_1)$	0.0003	1.880	5.270	40.22	0.4661
3	$s(b_2)$	0.0002	1.005	2.965	22.62	0.2622
4	$s(b_{11})$	8.323×10^{-7}	0.0054	0.0151	0.1150	0.0013
5	$s(b_{22})$	1.000×10^{-6}	0.0065	0.0181	0.1382	0.0016
6	$s(b_{12})$	7.680×10^{-7}	0.0050	0.0139	0.1061	0.0012

The values given in Table 4 were used for calculations of data presented in Table 5.

Table 5. Estimation of structural parameters relative to the error of estimation of the parameters.

No.	Indication of Estimate Error	Values of Errors in Determining				
		Boiler Energy Efficiency, $j = 1$	Emission of CO $j = 2$	Emission of NO_x $j = 3$	Emission of SO_2 $j = 4$	Emission of Dust $j = 5$
1	$\lvert b_0 \rvert / s(b_0)$	34.75	1.930	3.451	1.539	0.0128
2	$\lvert b_1 \rvert / s(b_1)$	2.000	2.446	2.556	1.237	0.2541
3	$\lvert b_2 \rvert / s(b_2)$	3.000	1.561	0.5376	1.176	1.692
4	$\lvert b_{11} \rvert / s(b_{11})$	1.304	2.335	2.365	1.130	0.00846
5	$\lvert b_{22} \rvert / s(b_{22})$	4.236	2.263	1.552	2.511	1.981
6	$\lvert b_{12} \rvert / s(b_{12})$	0.1141	0.2130	1.028	0.8079	0.6833

Comparing the values given in Table 5 with the critical value $t_c = 2.052$, it should be stated that structural parameters play an important role: $b_0^{(1)}$, $b_1^{(1)}$, $b_2^{(1)}$, $b_{22}^{(1)}$, $b_2^{(2)}$, $b_{11}^{(2)}$, $b_{22}^{(2)}$, $b_{12}^{(2)}$, $b_1^{(3)}$, $b_2^{(3)}$, $b_{11}^{(3)}$, $b_{22}^{(4)}$. For dust emissions, none of the structural parameters meets the materiality criteria.

3.3. Uncertainty of the Measured Quantity

In this analysis, most of measurement results used came from literature sources cited in Section 2.2. Knowledge of the uncertainty (class) of measuring instruments and test methods used is required for adequate assessment of the uncertainty of measurement results. The instrument class is not provided in the used literature sources. Taking into account the efficiency of boilers, it should be assumed that their energy efficiency was determined by an indirect method requiring the determination of individual energy losses. This requires determining the uncertainty of complex quantities requiring more measurements. No further information was provided. At the same time, it should be noted that ignorance of the uncertainty of the measured quantities is not required to carry out the tests described in Sections 3.2.1 and 3.2.2.

4. Discussion and Conclusions

Generalizations of determined characteristics should be considered a valuable and innovative element of this study by using 10 variables and independent sources to determine them. The experimental results, taken into account from various sources, concerned boilers in which hard coal was burned with different elemental compositions. The same large variation occurred in the case of biomass. The major differences concerned especially the composition of the mineral substance and the sulphur content. It should be assumed that the design and operating conditions of the boilers also differed.

The main achievement of the work is to determine the studied characteristics and determine the ranges of their usefulness.

The form of energy performance and ecological characteristics are set out in Table 2. The ranges of usability of these characteristics are varied. Energy performance can be used in the analysis of predictions in the field of technical and economic effects, especially useful when assessing the level of operation and modernisation of the existing dust steam boilers and the designing of new dust steam boilers with steam capacity from 130 to 230 t/h.

For the NO_x emission stream characteristics, its usefulness should be assessed as less than that of energy performance. This is mainly due to the greater than 0.2 value of the random variation coefficient v (see: Table 3). When assessing the usefulness of CO emission stream characteristics, it should be stated that the indicator v significantly exceeds the standard value of 0.2, but at the same time relatively favourable values are obtained when assessing structural parameter errors (see: Table 5). Therefore, it is proposed to use NO_x and CO characteristics for indicative analyses of boiler operation and design.

In turn, the determined characteristics of dust and SO_2 emissions do not meet the statistical standards required for determining analogous relationships. This is evidenced by both graphically illustrated discrepancies between the predicted and measured values of these emissions (Figures 5 and 6), as well as the corresponding values of the random coefficient of variation v, significantly higher than 0.2. Additional confirmation of this fact lies in errors in the estimation of structural parameters related to these emissions (see Table 5).

Explanations of the reasons for unsatisfactory effects of determining ecological characteristics are given in the notes provided in Figures 4 and 6.

At the same time, the analyses showed that energy efficiency of boilers is adequately characterized by their efficiency and chemical energy of the burned biomass.

However, it should be noted that the operating conditions occurring in the analyzed boilers can significantly affect the values obtained on the basis of the NO_x and CO emission flux characteristics. The forms of these characteristics determined in the paper relate to average operating conditions occurring in boilers of this type during measurements.

Author Contributions: Conceptualization, J.K., J.C. and M.K.; methodology, J.K., J.C. and M.K.; software, J.C. and P.Z.; validation, J.C. and M.K.; data analysis, J.C. and M.K.; writing—original draft preparation, J.K., J.C. and M.K.; writing—review and editing, J.K. and P.Z.; funding acquisition, P.Z.; supervision, J.K. All authors have read and agreed to the published version of the manuscript.

Funding: This research received no external funding.

Conflicts of Interest: The authors declare no conflict of interest.

References

1. Macuk, R.; Maćkowiak-Pandera, J.; Gawlikowska-Fryk, A.; Rubczyński, A. *Transformacja Energetyczna w Polsce—Edycja 2019*; Technical Report; Forum Energii: Warszawa, Poland, 2019.
2. European Commission. *Eurostat Database*; Eurostat: Kirchberg, Luxembourg, 2020.
3. GUS. *Statistics Poland. Energy from Renewable Sources in 2018*; Technical Report; GUS: Warszawa, Poland, 2020.
4. GUS. *Statistical Yearbook of the Republic of Poland 2019*; Technical Report; GUS: Warszawa, Poland, 2018.
5. Bartoszewicz-Burczy, H. Biomass potential and its energy utilization in the Central European countries. *Energetyka* **2012**, *12*, 860–866.
6. Wiśniewski, G. *Spalimy 10 Miliardów złotych. Współspalanie Wraca na Niespotykaną Skalę w Aukcjach 2020 na Energię z OZE*; CIRE: Warszawa, Poland, 2020.
7. IEA. *Coal Information: Overview*, 2019 ed.; Technical Report; IEA: Paris, France 2019.
8. IEA. *Coal 2019*; Technical Report; IEA: Paris, France, 2019.
9. Roni, M.S.; Chowdhury, S.; Mamun, S.; Marufuzzaman, M.; Lein, W.; Johnson, S. Biomass Co-Firing Technology with Policies, Challenges, and Opportunities: A Global Review. *Renew. Sustain. Energy Rev.* **2017**, *78*, 1089–1101. [CrossRef]
10. Madanayake, B.N.; Gan, S.; Eastwick, C.; Ng, H.K. Biomass as an Energy Source in Coal Co-Firing and Its Feasibility Enhancement via Pre-Treatment Techniques. *Fuel Process. Technol.* **2017**, *159*, 287–305. [CrossRef]

11. Cuellar, A.; Herzog, H. A Path Forward for Low Carbon Power from Biomass. *Energies* **2015**, *8*, 1701–1715. [CrossRef]

12. Kleinhans, U.; Wieland, C.; Frandsen, F.J.; Spliethoff, H. Ash Formation and Deposition in Coal and Biomass Fired Combustion Systems: Progress and Challenges in the Field of Ash Particle Sticking and Rebound Behavior. *Prog. Energy Combust. Sci.* **2018**, *68*, 65–168. [CrossRef]

13. Truong, A.H.; Patrizio, P.; Leduc, S.; Kraxner, F.; Ha-Duong, M. Reducing emissions of the fast growing Vietnamese coal sector: The chances offered by biomass co-firing. *J. Clean. Prod.* **2019**, *215*, 1301–1311. [CrossRef]

14. Milićević, A.; Belošević, S.; Crnomarković, N.; Tomanović, I.; Tucaković, D. Mathematical modelling and optimisation of lignite and wheat straw co-combustion in 350 MWe boiler furnace. *Appl. Energy* **2020**, *260*, 114206. [CrossRef]

15. Tan, P.; Ma, L.; Xia, J.; Fang, Q.; Zhang, C.; Chen, G. Co-firing sludge in a pulverized coal-fired utility boiler: Combustion characteristics and economic impacts. *Energy* **2017**, *119*, 392–399. [CrossRef]

16. Smith, J.S.; Safferman, S.I.; Saffron, C.M. Development and application of a decision support tool for biomass co-firing in existing coal-fired power plants. *J. Clean. Prod.* **2019**, *236*, 117375. [CrossRef]

17. Ko, S.; Lautala, P. Optimal Level of Woody Biomass Co-Firing with Coal Power Plant Considering Advanced Feedstock Logistics System. *Agriculture* **2018**, *8*, 74. [CrossRef]

18. Golec, T. Współspalanie biomasy w kotłach energetycznych. *Energetyka* **2004**, *7*, 11–24.

19. Sami, M.; Annamalai, K.; Wooldridge, M. Co-firing of coal and biomass fuel blends. *Prog. Energy Combust. Sci.* **2001**, *27*, 171–214. [CrossRef]

20. Gad, S.; Pawlak, A. Wpływ spalania biomasy na bezpieczeństwo procesu technologicznego w elektrowni konwencjonalnej. *Logistyka* **2014**, *3*, 1864–1870.

21. Steer, J.; Marsh, R.; Griffiths, A.; Malmgren, A.; Riley, G. Biomass co-firing trials on a down-fired utility boiler. *Energy Convers. Manag.* **2013**, *66*, 285–294. [CrossRef]

22. Xu, W.; Niu, Y.; Tan, H.; Wang, D.; Du, W.; Hui, S. A New Agro/Forestry Residues Co-Firing Model in a Large Pulverized Coal Furnace: Technical and Economic Assessments. *Energies* **2013**, *6*, 4377–4393. [CrossRef]

23. Kommalapati, R.; Hossan, I.; Botlaguduru, V.; Du, H.; Huque, Z. Life Cycle Environmental Impact of Biomass Co-Firing with Coal at a Power Plant in the Greater Houston Area. *Sustainability* **2018**, *10*, 2193. [CrossRef]

24. Yang, B.; Wei, Y.M.; Hou, Y.; Li, H.; Wang, P. Life cycle environmental impact assessment of fuel mix-based biomass co-firing plants with CO2 capture and storage. *Appl. Energy* **2019**, *252*, 113483. [CrossRef]

25. Zuwala, J.; Sciazko, M. Full-scale co-firing trial tests of sawdust and bio-waste in pulverized coal-fired 230t/h steam boiler. *Biomass Bioenergy* **2010**, *34*, 1165–1174. [CrossRef]

26. Zheng, S.; Yang, Y.; Li, X.; Liu, H.; Yan, W.; Sui, R.; Lu, Q. Temperature and emissivity measurements from combustion of pine wood, rice husk and fir wood using flame emission spectrum. *Fuel Process. Technol.* **2020**, *204*. [CrossRef]

27. Mehmood, S.; Reddy, B.V.; Rosen, M.A. Energy Analysis of a Biomass Co-firing Based Pulverized Coal Power Generation System. *Sustainability* **2012**, *4*, 462–490. [CrossRef]

28. Polański, Z. *Planowanie Doświadczeń w Technice*; PWN: Warszawa, Poland, 1984.

29. Montgomery, D. *Design and Analysis of Experiments*; John Wiley & Sons Inc.: New York, NY, USA, 2001.

30. Myers, R.; Montgomery, D. *Response Surface Methodology*; Wiley & Sons Inc.: New York, NY, USA, 1995.

31. StatSoft. *Elektroniczny Podręcznik Statystyki PL*; StatSoft: Warszawa, Poland, 2006.

32. Kozioł, J.; Czubala, J. An optimisation strategy using probabilistic and heuristic input data for fuel feeding boilers with regard to the trading effects of CO2 allowances. *Energy* **2013**, *62*, 82–87. [CrossRef]

33. Koziol, J.; Koziol, M. Determining operating characteristics of co-firing processes in grate furnaces. *Fuel* **2019**, *258*, 116164. [CrossRef]

34. Mann, M.; Spath, P. A life cycle assessment of biomass cofiring in a coal-fired power plant. *Clean Prod. Processes* **2001**, *3*, 81–91. [CrossRef]

35. Canalís, P.; Royo, J.; Sebastián, F. Influence of co-combustion in the efficiency of a pulverized coal boiler. In Proceedings of the 14th European Biomass Conference, Paris, France, 17–21 October 2005.

36. EPA. *Biomass Combined Heat and Power Catalog of Technologies*; Technical Report; EPA: Washington, DC, USA, 2007.

37. Valero, A.; Canalís, P.; Palacio, J.; Pascual, J.; Royo, J.; Sebastián, F.; Tapia, R. *Co-Firing of Low Rank Coal and Biomass: A Chance For Biomass Penetration in the Renewables*; Technical Report; Center of Research for Power Plant Efficiency: Zaragoza, Spain, 2020.

38. Ściążko, M.; Zuwała, J.; Pronobis, M. Zalety i wady współspalania biomasy w kotłach energetycznych na tle doświadczeń eksploatacyjnych pierwszego roku współspalania biomasy na skalę przemysłową. *Energetyka i Ekologia* **2006**, *3*, 207–220.

39. Nicholls, D.; Zerbe, J. Cofiring biomass and coal for fossil fuel reduction and other benefits-status of North American facilities in 2010. *USDA For. Serv. Gen. Tech. Rep. PNW-GTR* **2012**, *867*, 1–22. [CrossRef]

40. Zuwała, J. Wpływ współspalania biomasy z paliwami konwencjonalnymi na parametry eksploatacyjne pracy bloków energetycznych. *Energetyka* **2010**, *2*, 108–114.

41. Oehlert, G.W. *A First Course in Design and Analysis of Experiments*; Technical Report; University of Minnesota: Minneapolis, MN, USA, 2010.

42. Bartosiewicz, S. *Estymacja Modeli Ekonometrycznych*; PWE: Warszawa, Poland, 1989.

43. Sobczyk, M. *Statystyka*; PWN: Warszawa, Poland, 1996.

44. Welfe, W.; Welfe, A. *Ekonometria Stosowana*; PWE: Warszawa, Poland, 2004.

Article

Improving Municipal Solid Waste Management Strategies of Montréal (Canada) Using Life Cycle Assessment and Optimization of Technology Options

Tahereh Malmir, Saeed Ranjbar and Ursula Eicker *

Canada Excellence Research Chair Next Generation Cities, Gina Cody School of Engineering and Computer Science, Concordia University, Montréal, QC H3G 1M8, Canada; tahereh.malmir@mail.concordia.ca (T.M.); saeed.ranjbar@mail.concordia.ca (S.R.)
* Correspondence: ursula.eicker@concordia.ca

Received: 15 October 2020; Accepted: 29 October 2020; Published: 31 October 2020

Abstract: Landfilling of organic waste is still the predominant waste management method in Canada. Data collection and analysis of the waste were done for the case study city of Montréal in Canada. A life cycle assessment was carried out for the current and proposed waste management system using the IWM-2 software. Using life cycle assessment results, a non-dominated sorting genetic algorithm was used to optimize the waste flows. The optimization showed that the current recovery ratio of organic waste of 23% in 2017 could be increased to 100% recovery of food waste. Also, recycling could be doubled, and landfilling halved. The objective functions were minimizing the total energy consumption and CO_{2eq} emissions as well as the total cost in the waste management system. By using a three-objective optimization algorithm, the optimized waste flow for Montréal results in 2% of waste (14.7 kt) to anaerobic digestion (AD), 7% (66.3 kt) to compost, 32% (295 kt) to recycling, 1% (8.5 kt) to incineration, and 58% (543 kt) to landfill.

Keywords: organic waste; energy recovery; life cycle assessment; cost analysis

---

1. Introduction

The increased population in urban areas leads to a significant raise in waste generation. Conventional waste management methods like landfill are amongst the main contributors to greenhouse gas (GHG) emissions in the world. Landfills occupy large areas of lands and potentially pose a risk to human health and the surrounding environment. Therefore, it is crucial to find alternative methods for better municipal solid waste management (MSWM) in urban areas. Annually, 2.01 billion tons of waste are produced globally, and waste to energy technologies provide approximately 1.5% of the final energy consumption in Europe [1].

Efforts have been made worldwide to reduce landfilling of the biodegradable fraction of waste, but the reduction amount is not satisfying so far. An example is food wastes (FW), which are the easily biodegradable organic waste (OW) [2]. It contributes almost half of the total municipal wastes in most countries [3] and has great potential to be used for energy purposes. However, it is directly landfilled in many cases. In 2014 and 2015, FW accounted for 38 [4] and 39 million tons [5] in the USA, respectively, and three-quarters of these amounts were landfilled [4,5]. The EU Landfill Directive in 1999 prevented landfilling the OW and forced the members to reduce the quantity of biodegradable municipal waste sent to landfills to 75% (2006), 50% (2009), and then 35% (2016) compared to 1995 [6]. Based on this directive, the proportion of municipal waste disposed of by landfilling should be reduced to 10 % or less of the total amount of municipal waste generated by 2035 but most of the European countries could not achieve this target [7]. In 2018, 247 million tons of municipal solid waste (MSW) were treated in the EU by landfilling (23%), recycling (30%), composting (17%) and incineration with or without energy recovery (47%) [8].

Life cycle assessment (LCA) evaluates the potential environmental impacts of products. IWM-2 is a life cycle inventory (LCI) model for integrated waste management that predicts the environmental burdens of a specific waste management system [9].

Numerous studies have focused on the LCA of diverse waste management systems. [1] compared recovery methods, which are beneficial compared to disposal options and concluded that thermal treatment and anaerobic digestion (AD) might be favorable over composting. Composting usually require large areas, is highly affected by weather condition and has odor problems. [10] criticized large-scale centralized composting due to enhanced environmental impacts and therefore, considered it a temporary solution. Decentralized waste management systems decrease transportation requirements significantly. [11] found out that the presence of a composting plant at 10 km from the municipality would decrease 65% of the environmental impacts due to the external transport.

Connecting agriculture and waste is beneficial in terms of the reduction of GHG emissions. Still, there are some challenges regarding increased costs and acceptance for the use of digestate as a fertilizer (e.g., legal restrictions on the use of digestate produced from sewage sludge) [12,13] reported that N_2O emissions of the application of liquid and solid portions of the digestate were the most significant contributors to global warming among all the life cycle stages. In another study, [14] coupled AD with composting to reduce the drawbacks associated with the direct soil distribution of anaerobic digestates such as the emission of CO_2 and obtained stable products to be safely used in soils without affecting their N- and P-fertilizing capacity.

According to [15], one way to improve the utilization efficiency of biomass is the use of waste and production residues, and a vast majority of waste to energy technologies have lower GHG emissions when compared to fossil fuels.

The performance of waste management methods depends on the waste composition and climate conditions. The agglomeration of Montréal (MTL) in the province of Québec (Canada) was chosen as the case study. Due to the severe weather condition in this province, the energy consumption of Québec's residents is one of the highest in the whole world. Currently, about half of the energy demand in Québec is supplied through renewable sources and the Québec government has ambitious plans of increasing this amount to 60.9% in 2030 [16]. One of the key targets in this plan is increasing bioenergy production by 50% by 2030. MTL as the biggest city in this province plays a vital role in achieving those targets. Therefore, in this study the authors focused on the current waste flow in MTL and presented different possible scenarios for MSWM in this city. The challenges to achieve optimized waste flows are discussed. The focus of the present study is on municipal solid waste, which includes residential, commercial, and institutional waste, and excludes industrial, construction, and hazardous waste [17]. To the best of authors' knowledge, there are no LCA studies of waste, and especially OW management for the chosen case study city.

The paper is organized as followed: First, the waste composition in MTL, and the current waste flow in the city is presented. Then, different MSWM scenarios are defined and their environmental performance are compared using IWM-2 LCA methodology. In the next section, based on the preliminary LCA results, a mathematical model is developed and the waste flow is optimized to minimize the equivalent CO_2 of GHG emissions, total energy consumption in waste management system, and total cost of the system. In the final section, the challenges in the way to reach to this optimized waste flow are discussed. The aim of this study was LCA of the current waste management system and waste management systems with new technologies in Montréal.

2. Current Status of Waste Management in Montreal

The work uses the Canadian city Montréal as a case study. The agglomeration of Montréal (MTL) includes 16 cities together with the City of Montréal, which in turn is divided into 19 boroughs. The City of Montréal is the largest city in the Canadian province of Québec (24% of the population) [18] and the second-most populous municipality in Canada, with around 2 million inhabitants [19,20]. Currently, most of the OW in Québec is landfilled or combusted, and it is planned to ban the disposal of OW and reach 60% diversion from landfill [21]. Moreover, MTL has a Waste Management Master Plan firmly anchored in the targets of the Québec Residual Materials Management Policy with its 2011–2015 Action Plan of the Government of Québec. According to this plan, the recovery target for recyclables, OW, and construction and demolition waste (CD) is 70%, 60%, and 70%, respectively [19]. It is also planned, by 2030, to increase bioenergy production by 50% through various methods such as bio-methanation of OW [16]. Mishandling of OW in MTL or Québec has been reported by several studies [18,22]. Therefore, the environmental assessment of the current and proposed waste management systems is essential regarding their impact on energy consumption and emissions.

2.1. Data Collection and Analysis

Data collection and analysis for the waste flow of MTL was mainly done based on the reports published by the Service de l'Environnement in MTL [19,23]. Their report characterizes waste collections from three sources, including residential, institutional, and commercial ones. The generated waste included OW (FW, yard waste and mixed residue), paper and cardboard (PC), metal, glass, plastic (MGP), CD, harmful household products (, paint, pesticide, mercury devices, etc.), textile, E-waste, and household waste. Collection and disposal of waste are handled by the municipalities in MTL in different ways, and separation of materials is done in sorting centers. Curbside collection service collects the household waste and recyclables, and partially OW. OW consisting of FW and yard waste is collected in most of the buildings of eight or fewer dwellings in MTL, and then transformed into compost. Seven eco-centers in MTL collect CD, wood, metal, tire, polystyrene and textile, harmful household products, E-waste, yard waste (gardening and weeding residues, leaves, and Christmas trees), and other reusable materials. CD is also collected on the street or as a result of resident calls. Household waste and non-recyclable CDs are sent to the landfills.

2.2. Trend in Waste Generation

Figure 1 illustrates the generated waste in MTL (2012 to 2017). In MTL, total waste generation decreased from 970 kt in 2012 to 931 kt in 2017. The average amount of OW, MGP and PC, CD, and textile, E-waste & harmful household products was 361 kt, 286 kt, 234 kt, and 8 kt, respectively. Various factors affected the decrease in waste quantities such as replacement of printed newspapers by digital editions, eco-design of products which reduces the weight of containers, reduction of over-packaging, and reduction of consumption.

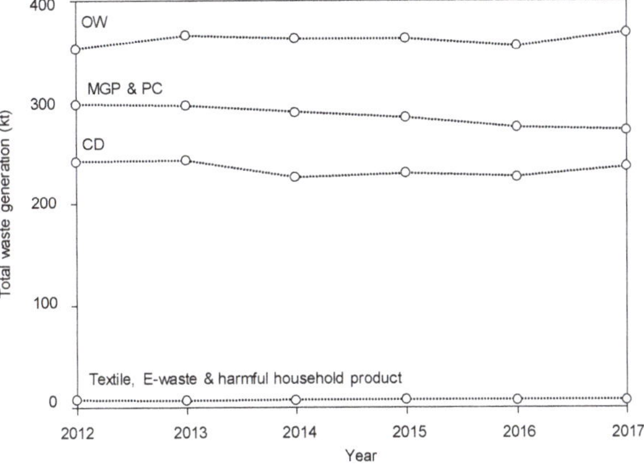

Figure 1. Total waste generation in MTL [19,23]; OW: organic waste, PC: paper and cardboard, MGP: metal, glass, plastic, CD: construction and demolition waste, and E-waste: electronic products.

Many products that used to be made from recyclable materials changed to multi-layered flexible packaging that are not accepted for recycling in MTL. For example, in MTL, plastic #6 (polystyrene), different kinds of plastic bags and films are not considered as recyclable items. The number of public policies on plastic carrier bags has more than tripled since 2010, and they are now found on all continents, ranging from the municipal to the intergovernmental level [24]. Introducing the degradable plastics as the environmental-friendly alternatives to the market can decrease the vast amounts of plastics that are landfilled. For instance, Malmir used a solvent casting method to prepare biodegradable films of poly (3-hydroxybutyrate-co-3-hydroxyvalerate) with cellulose nanocrystals, which has the capability for applications in the industry of food packaging [25] and achieved well-dispersed bionanocomposites with improved mechanical and barrier properties [26].

2.3. Waste Flow

The waste flow of MTL in 2017 is shown in Figure 2. Based on this figure, in MTL, 931 kt waste was generated and the whole amount of this waste was the subject of the research. This amount was comprised of 95% recyclables (OW, PC, MGP, CD, textile, E-waste, and harmful household product) and 5% non-recyclables (non-recyclable CD and other materials), and the portion of diverted and landfilled waste was 45% and 55%, respectively. Adding mixed paper from OW (139 kt) to PC and MGP (272 kt), recyclables could account for 357 kt. OW in MTL accounted for 369 kt from which around 85 kt was recovered. The recovery ratio of OW increased from 11% in 2012 to 23% in 2017 but was still far from the 60% recovery target in 2011–2015 Action Plan of the Government of Québec. The recovery ratio of PC and MGP, and CD was 60%, and 68%, respectively. To compare, household waste collected from urban and rural sectors of Saguenay in the Canadian province of Québec comprised of 53% to 66% OW, 4% PC, 15% MGP, and 5% textile [27].

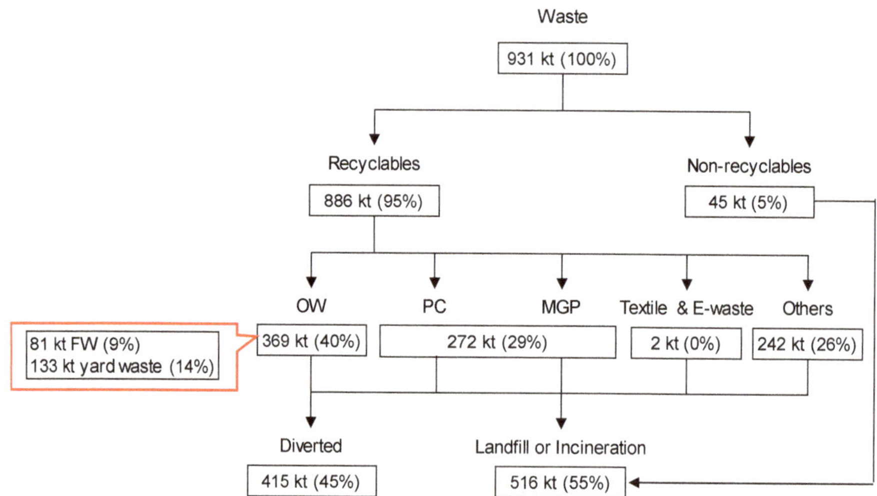

Figure 2. Waste flow of MTL in 2017 [19,23].

The waste flow also shows the percentage of FW and yard waste for the OW of MTL. This percentage was not available for 2017, and we assumed the same percentage in 2016 [19]. Accordingly, FW accounted for 9% (81 kt), and yard waste was 14% (133 kt). The rest of the OW was 17% (155 kt) mixed residue.

3. Methodology

3.1. LCA Methodology

IWM-2 was used as the LCA methodology to predict the environmental burdens of integrated waste management systems [28,29]. The scope of the environmental analysis model was defined to include the major components of residential waste, including paper, plastic, glass, aluminum and steel, FW, and yard waste. Other types of wastes were considered as components which could be treated through energy recovery and landfilling options [29]. Goals, functional unit, and system boundary, life cycle inventory, and life cycle impact assessment in the following sections were based on IWM-2.

3.1.1. Goals, Functional Unit and System Boundary

LCA of the current waste management systems and waste management systems with new technologies in MTL was considered. These technologies were based on composting of FW, and energy recovery from mainly FW and yard waste. The total waste generated in MTL in 2017 was considered as a functional unit in the mentioned systems. The model evaluated the environmental burdens associated with waste management from the point at which a material is discarded into the waste stream to the point at which it was either converted into a useful material or, it was finally disposed [28]. Accordingly, waste collection, waste transfer, sorting of recyclable materials at a materials recovery facility, reprocessing of recovered materials into recycled materials, composting, energy recovery and landfilling were evaluated by the model through recycling of paper, plastics, glass, steel, and aluminum, composting and AD of paper, yard waste and food waste, and incineration and landfilling of all waste components [29]. However, in this study, only AD and composting of FW and incineration of yard waste was considered.

3.1.2. Life Cycle Inventory

The analysis of all the material and energy inputs and outputs for each stage in the life cycle could be combined to give the overall life cycle inventory [9]. The overall estimation of energy consumption and emissions of the waste management systems in this study was conducted with the help of the IWM-2 model and its pre-defined standard data in Microsoft Excel for Office 365 MSO version 16 and Microsoft Visual Basic for Applications version 2012.

3.1.3. Life Cycle Impact Assessment

The model estimated the energy consumed (or produced) and the emissions to air, water, and land associated with different waste management practices [28]. The specific indicator parameters evaluated, and the environmental effects associated with these parameters are shown in Table 1.

3.2. Waste Management Scenarios

LCA has been conducted for the current waste management systems in MTL (Sc1) and three proposed scenarios (Sc2, Sc3, and Sc4) in which the energy consumption and emissions have been determined. All the proposed scenarios considered the maximum amount of recycling rates. They also fed all the yard waste to incineration technology because lignin does not undergo AD, and cellulose and hemicellulose are degraded slowly in comparison [1]. In the case of FW, Sc2 specified all the FW for the AD technology, Sc3 assumed all the FW for the composting technology, and Sc4 allocated half of the FW for the AD technology and the other half for the composting technology. Figure 3 shows the amount of input waste for waste management scenarios in MTL.

Table 1. Indicator parameters [28,30].

Indicator Parameter	Indicator of	Unit
Energy Total energy consumed	Resource depletion	GJ
Emissions to air Greenhouse gases Carbon dioxide (CO_2) Methane (CH_4)	Climate change	t
Acid Gases Nitrogen oxides (NOx) Sulphur dioxide (SO_2) Hydrogen chloride (HCl)	Acidification, health risk	t
Smog precursors Volatile organic compounds (VOC) Nitrogen oxides (NOx) Particulate matter (<10 microns) (PM-10)	Urban smog formation, health risk	t
Heavy metals Lead (pb) Cadmium (Cd) Mercury (Hg)	Health risk	kg
Trace organics Dioxins & Furans (TEQ)	Health risk	g
Emissions to water Heavy metals Lead (pb) Cadmium (Cd) Mercury (Hg)	Health risk, environmental degradation	kg
Trace organics Dioxins & Furans (TEQ)	Health risk, environmental degradation	mg
Biochemical Oxygen Demand (BOD)	Water quality, environmental degradation	kg
Emissions to land Residual solid waste	Land use disruption	t

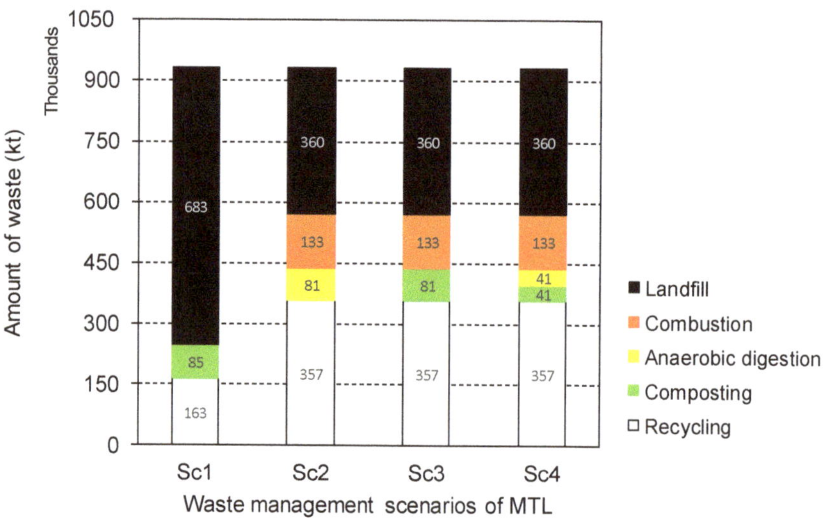

Figure 3. Amount of input waste for waste management scenarios in MTL.

3.3. Optimization

LCA results are useful measures to develop a mathematical model for the energy and environmental performance of a waste management system. The IWM-2 software is a helpful tool for conducting LCA for a waste management system as it includes all different parts of the system, including transportation, sorting, and energy recovery. The results obtained from IWM-2 included the energy consumption and the CO_2 equivalent of GHG emissions from each waste management technology that were good indicators of waste management system performance. Based on the results derived from the different proposed scenarios, the ranges of the waste sent to each technology was defined. According to these values, the amount of waste sent to each section was changed to obtain the equivalent CO_2 of GHG emissions (CO_{2eq}) and energy consumption of each technology (E) for the specified waste amount. By using these results, a curve fitting tool was applied to develop a second-order mathematical relationship for both energy consumption and CO_{2eq} of GHG as a function of waste input [30]. To achieve a better fit, the data for all technologies were normalized. The general form of the Equation for each technology is as follows:

$$E = a\,[(x - \mu)/\sigma]^2 + b\,[(x - \mu)/\sigma] + c, \tag{1}$$

$$CO_{2eq} = a\,[(x - \mu)/\sigma]^2 + b\,[(x - \mu)/\sigma] + c. \tag{2}$$

where, E is the energy consumption in each technology, CO_{2eq} is the equivalent CO_2 of GHG emissions from each technology, x is the amount of waste sent to each section, μ and σ are the mean and variance of the data obtained for each technology, respectively, and a and b and c are constants. The values of μ and σ were derived from the curve fitting tool utilized for developing the Equations.

3.3.1. Objective Functions

The total energy consumption and CO_{2eq} emission were calculated from the following equations [30]:

$$E_{total} = E_{AD} + E_C + E_{Cbs} + E_L + E_R, \tag{3}$$

$$CO_{2total} = CO_{2AD} + CO_{2C} + CO_{2Cbs} + CO_{2L} + CO_{2R}. \tag{4}$$

In Equation (3), E_{AD}, E_C, E_{Cbs}, E_L, and E_R are energy consumption at AD, composting, incineration, landfilling, and recycling units, respectively. In the energy from waste facilities including AD and incineration, the energy consumption was calculated from the electrical energy generated minus the energy consumed. For the recycling unit, the saved energy by using recovered material was subtracted from the energy consumed for recycling the material.

In Equation (4), CO_{2AD}, CO_{2C}, CO_{2Cbs}, CO_{2L}, and CO_{2R} are GHG emissions of the anaerobic digestion, composting, incineration, landfilling, and recycling units, respectively. These quantities show the total emissions of each technology in their life cycle based on equivalent CO_2.

These two quantities are functions of waste input in each technology. In this optimization procedure, two of the objective functions were minimizing the total energy consumption and CO_2 equivalent of GHG emissions.

Cost is a vital factor in designing an integrated waste management system. Table 2 contains the estimated cost function for each waste management technology. The costs were categorized into the initial capital costs and operating costs and the parameter x denotes the annual waste input of each technology.

Table 2. Costs in Canadian dollars per ton of annual waste input.

Technology	Capital Cost (CAD/t)	Operating Costs (CAD/t)	Reference
Recycling	$190\,x$	$190\,x$	[31]
Composting	$4000\,x^{0.7}$	$7000\,x^{-0.6}$	[32]
AD	$35000\,x^{0.6}$	$17000\,x^{-0.6}$	[32]
Incineration	$5000\,x^{0.8}$	$700\,x^{-0.3}$	[32]
Landfilling	$6000\,x^{0.6}$	$100\,x^{-0.3}$	[32]

In addition to the total energy consumption and CO_{2eq} emission (Equation (3) and Equation (4)), the total cost of the waste management system was also considered as an objective function to be minimized [30]:

$$Cost_{total} = Cost_{2AD} + Cost_{2C} + Cost_{2Cbs} + Cost_{2L} + Cost_{2R}. \tag{5}$$

Consider X to be a vector containing the waste input of each technology (X= $(x_{AD}, x_C, x_I, x_L, x_R)$), and the arrays of this vector were decided by the optimization constraints, then X* was optimal in the space S, if energy, CO_2 and total cost are minimized:

$$E_{total}(X^*) \;\&\; CO_{2eq.total}(X^*) \;\&\; Cost_{total}(X^*) \leq E_{total}(X) \;\&\; CO_{2eq.total}(X) \;\&\; Cost_{total}(X) \text{ for all } X \in S \tag{6}$$

3.3.2. Constraints

Based on the results obtained from IWM-2, the best scenario was chosen. Therefore, the optimization constraints could be determined according to the chosen scenario. The lower bounds for the waste input of landfill and recycling units were the current amount of waste sent to these units in MTL.

3.3.3. Method

An optimization algorithm was used to find the best waste flow for the waste management system in MTL. This optimization algorithm was a multi-objective one as the proposed system should be both environmentally friendly and economically feasible. GA is a popular option for solving such constrained multi-objective optimization problems. GA has been evolved into different forms that each of them is different from the original GA. One of these evolved forms is a non-dominated sorting genetic algorithm (NSGA) developed by Srinivas and Deb [33]. The difference between NSGA and original GA is only in how the selection operator works while crossover and mutation operators remain the same.

In this study, an improved form of NSGA, meaning NSGA-II [34] was used to minimize the energy consumption, CO_{2eq} of GHG emissions, and cost of the system. This improved algorithm was less complicated in terms of calculations, and the solutions were more diverse compared to original NSGA [35].

Figure 4 illustrates the procedure of optimization. The first step was an initialization, which included defining objective functions, input variables, and constraints. The mass input of the five technologies (AD, compost, incineration, landfill, and recycling) were taken as input variables. In the next step, a set of values was assigned to the defined input variables as the initial population and a fitness function was found for each set of answers in the next step. Then, using this initial population, the values of the objective functions were calculated to identify the answers that minimize them. Next, the non-dominated sorting was done to order the answers based on their fitness functions. In step 4, parent chromosomes were chosen among the ordered initial population. The crossover process was used to generate children for the chosen parents. In the following step, the mutation operator was utilized for the children. Unlike the crossover process where the children have the same characteristics as the parents, after mutation, some of the children gain characteristics that belong to neither of parents. Then, these mutated children are mixed with other children, and again non-dominated sorting will occur, and children are chosen for the next generation. Finally, the stop criterion of the algorithm was checked. The steps 2 to 8 were repeated until this criterion is met.

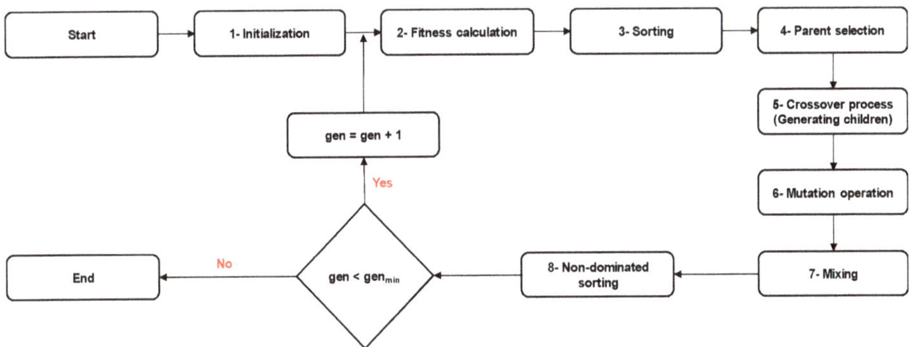

Figure 4. Procedure of optimization.

4. Results and Discussion

4.1. LCA Results

Greenhouse gases (carbon dioxide (CO_2), methane (CH_4) and nitrogen oxides (NOx), and the CO_{2eq} in kt of CO_2), acid gases (nitrogen oxides (NOx), sulfur oxides (SOx) and hydrochloric acid (HCl)), smog emissions (NOx, particles (PM) and volatile organic compounds (VOC)), energy consumption and remaining amounts in the current status of waste management and the proposed scenarios in MTL are presented in Table 3. In MTL, Sc1 consumes the most energy (6869 TJ saving) and emits the least CO_{2eq} (69 kt). Sc3 and Sc4 are the better waste management scenarios in which 14027 TJ and 14043 TJ energy is saved, and 127 kt and 144 kt CO_{2eq} is emitted, respectively. Each waste management system has its own residuals, for example ash in incineration, or residuals that could not being further recycled in recycling etc. These residuals are usually sent to landfill.

Table 3. The emissions in the current status and proposed scenarios of waste management in MTL.

Case	GHG Emissions (kt)			Acid Gases Emissions (kt)			Smog Emissions (kt)			Residuals (kt)	Energy (TJ)
	CO_2	$CH_4 + NO_X$	CO_{2eq}	NO_X	SO_X	HCl	NO_X	PM	VOC		
Sc1	−76	6	69	0.07	0.02	0.001	0.07	0.22	0.05	695	−6869
Sc2	9	4	142	0.19	0.04	0.02	0.19	0.13	0.05	427	−13,530
Sc3	9	3	127	0.17	0.04	0.02	0.17	0.15	0.04	417	−14,027
Sc4	11	3	144	0.20	0.04	0.02	0.20	0.14	0.06	423	−14,043

With Sc1 as the reference scenario of today's waste management system, Sc2 uses anaerobic digestion, Sc3 uses compost, and Sc4 uses anaerobic digestion and compost. The equivalent CO_2 is calculated using $CO_{2eq} = CO_2 + 21 * CH_4 + 310 * NOx$.

In the AD module (Sc2), emissions originate from biogas combustion (GHG contributor), aerobic composting (GHG contributor), and water (leachate) from the process. All the methane produced is typically combusted, and the resultant CO_2 emissions are not counted. Biogas combustion emits NO_X too, which contributes to equivalent CO_2 emission. Therefore, emissions of $CH_4 + NOx$ and equivalent CO_2 are higher for the scenarios with a higher AD ratio. The model does not consider offsetting the combustion of fossil fuels rather than biogas and, consequently, emission saving.

Process emissions of composting include only one direct emission, which originates from aerobic composting. GHG emissions of composting are thus lower than AD;

Among the proposed alternative scenarios, Sc2 indicates that using the AD unit as the sole FW treatment technology has the lowest efficiency in MTL. One of the reasons can be attributed to the

FW quantity in MTL which does not seem high enough for the maximum simultaneous reduction of energy consumption and CO_2 production by AD. It is worth mentioning that IWM-2 estimates the material-specific AD yields of CH_4 and CO_2 based upon the lab studies of AD of MSW in landfills [25]. Accordingly, various kinds of AD set-ups are neglected, and hence, further LCA studies are required based on lower FW quantity feeding to AD units in order to find out the energy consumption and CO2 emissions of each set-up. Studies showed that although there were 688 centralized AD plants for biowaste treatment (average capacity 31,700 ton/year) in EU in 2016, small scale AD (5.2 ton/year) can be technically viable with potential biogas production performance like large scale AD (3372 ton/year) [36].

4.2. Optimization Results

Based on the results obtained from IWM-2, the best scenario is when all the FW is divided between AD and compost, all the yard waste is sent to an incinerator, and all the recyclable materials are recovered. Therefore, the optimization constraints can be determined. As mentioned previously, the lower bounds for the waste input of landfill and recycling units are the current amount of waste sent to these units in MTL:

$$0 \leq x_{AD} \leq 81, \tag{7}$$

$$0 \leq x_C \leq 81, \tag{8}$$

$$x_{AD} + x_C = 81, \tag{9}$$

$$0 \leq x_{Cbs} \leq 133, \tag{10}$$

$$360 \leq x_L \leq 683, \tag{11}$$

$$163 \leq x_R \leq 357. \tag{12}$$

Table 4 shows the parameters of Equations (1) and (2) for each waste management technology, obtained by using MATLAB curve fitting tools. To explain more, based on the result of the fourth LCA scenario (Sc4), the waste input to each technology was changed and the amount of CO_{2eq} of GHG emissions and energy consumption of each technology was recorded. Then, two curves were fitted for unit based on their waste input (one for CO_2 and one for Energy). These values were used for doing the optimization. The results of two-objective and three-objective optimizations are presented in the following sections.

Table 4. Equation parameters for different waste management technologies.

Technology	Input		μ	σ	a	b	c
Recycling	x_R	Energy	2.6×10^5	6.2×10^4	-115	-1.2×10^6	-1.1×10^7
		CO_2			-10	-1.1×10^5	-9.4×10^5
Composting	x_C	Energy	1.3×10^4	1.4×10^4	23	1658	3.3×10^4
		CO_2			-0.03	160	8104
AD	x_{AD}	Energy	5.9×10^4	1.4×10^4	22	-5×10^4	-1.6×10^5
		CO_2			6	712	$1.8e4$
Incineration	x_{Cbs}	Energy	6.6×10^4	4.3×10^4	-3.6×10^4	-3×10^5	-3.8×10^5
		CO_2			284	7989	1.2×10^4
Landfilling	x_L	Energy	5.2×10^5	1×10^5	-243	3025	3.1×10^4
		CO_2			-974	8597	9.6×10^4

This three-objective optimization for the integrated waste management system of MTL considers the cost as an objective function in addition to total energy consumption and CO_{2eq} emissions of GHG. Table 5 shows the best three optimum waste flows for MTL.

Table 5. The best three waste flows resulted from optimization.

Design Number	x_{AD} (t)	x_C (t)	x_{Cbs} (t)	x_L (t)	x_R (t)	C_{total} (kt of CO_2)	$Cost_{total}$ (CAD)	E_{total} (Tj)
1	14,710	66,393	8580	54,0166	29,4955	−879	12,3000607	−11,809
2	14,525	66,393	7880	55,6522	29,4861	−877	12,3203356	−11,802
3	14,684	66,393	8271	54,2802	29,7906	−884	12,3464210	−11,866

In the best optimum design (design number 1) which has the lowest total cost among these three options based on the value derived for the total amount of waste, 58% of total waste should be sent to the landfill unit (540 kt). Recycling unit has the highest share after landfill (32%) equal to 295 kt. Taking FW into account, 14.7 kt should be sent to AD and 66.3 kt to composting unit, which accounts for 2% and 7% of the total waste, respectively. The incineration unit has the lowest share in the system (1% equal to 8.5 kt) of total waste (Figure 5).

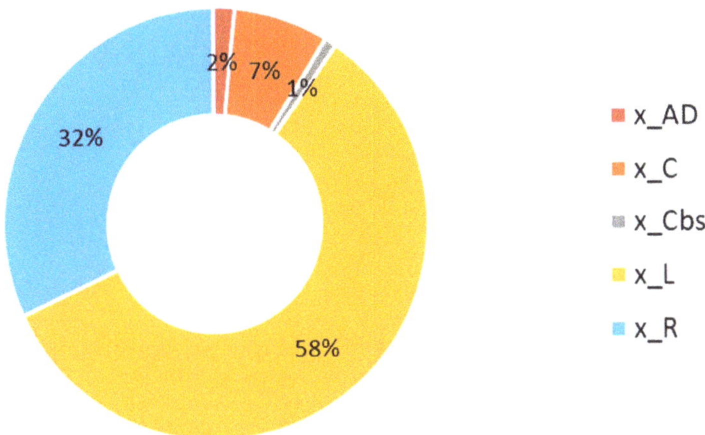

Figure 5. The optimized waste flow for MTL.

4.3. Discussion

It is clear from the results that the main benefit of having an integrated MSWM system is a significant reduction of energy consumption and emissions. LCA provides a comprehensive, consistent and transparent overview of flows in the waste management systems with quantification of the environmental profile [37]. Based on the optimization results, the amount of recycled waste in MTL should increase by 87 kt per year. Also, adding incineration and AD units to the waste management system of the city would increase the share of energy produced from renewable sources. The results are consistent with the study by [38] with increasing focus on recycling. Also, [39] assessed the environmental and economic benefit of the substitution of energy crops with food waste in AD and concluded that a reduction of 42% in the carbon footprint of the electricity produced from the biogas plant can be obtained. Moreover, installing new units like AD and incineration creates more jobs in the city which is a social benefit of this MSWM system. A study by [40] showed that new jobs could be created in the various processing centers and between transportation nodes of the waste management system. Employment opportunities by waste to energy include the collection and sorting of waste, waste transportation, waste plant construction, and plant operation. On average, a Waste-to-Energy plant in Europe can create 62 direct jobs, and the total direct and indirect jobs in 2011 was 56,000 in Europe [41]. However, it should be noted that all these results have been driven based on this assumption that all types of wastes are separated completely which is not possible in real life. The following bullet points summarizes different challenges present in the way to achieve this optimized waste flow:

(1)   As mentioned above, the biggest challenge is complete separation and sorting of different types of the waste. Creating sorting units for the whole amount of waste is an expensive solution. Therefore, another solution is encouraging people to be more interested in source separation of waste.

(2)   Hydro Québec is the only supplier of electricity in the province of Québec. Unfortunately, there is no specific policy about buying self-generated electricity from private suppliers. Therefore, it affects the interest from external investors to help to construct expensive units like AD and incineration.

(3)   The other existing challenge is the public awareness. People should become aware of the hazards of landfilling the municipal waste and realize what an important role they play in different waste management scenarios.

(4)   Another challenge is the location of new AD and incineration units and whether there should be one central unit or several distributed units across the city. Although a study by [36] concluded the advantages of a fully decentralized AD systems, the authors believe that more detailed LCA studies are needed to find the solution for this problem.

(5)   The severe weather condition in Montréal during its long winters is another challenge for utilizing organic waste management facilities like AD and composting units. Further thermal energy would be required to keep the system in an optimum temperature condition. Especially in case of AD and composting, cold weather might slow down the degradation process. Putting the AD in a greenhouse has been suggested and is recommended. Study by [42] showed that an AD could 49% less heat energy by being housed in a greenhouse.

5. Conclusions

The work presents an analysis of waste flow including OW, PC, MGP, CD, textile, E-waste, harmful household product, and other materials in a case study city Montréal. It shows the huge potential of energy recovery from FW and yard waste instead of landfilling them, as is the current OW management method.

In MTL in 2017, 931 kt waste was generated, and the portion of diverted and landfilled waste was 45% and 55%, respectively. OW in MTL accounted for 369 kt from which around 85 kt was recovered.

Four scenarios were analyzed to assess the greenhouse gas emissions and costs of different waste management strategies. With the current waste management system as the reference scenario 1, the proposed scenarios Sc2, Sc3, and Sc4 feed all the yard waste to incineration. Moreover, all the food waste goes to anaerobic digestion in Sc2 and to composting in Sc3. Sc4 considers 50% of FW for AD and composting.

The LCA study showed that in MTL, Sc3 and Sc4 are the better waste management scenarios in which 14,027 TJ and 14,043 TJ energy is saved, and 127 kt and 144 kt CO_{2eq} are emitted, respectively.

Based on the results obtained from LCA studies, NSGA-II was used as an optimization algorithm to optimize the waste flow in MTL. The objective functions were minimizing the total energy consumption and CO_{2eq} emission of GHG and the total cost in the waste management system. The optimized waste flow for MTL by using a three-objective optimization algorithm is sending 2% of waste (14.7 kt) to AD, 7% (66.3 kt) to compost, 32% (295 kt) to recycling, 1% (8.5 kt) to incineration, and 58% (543 kt) to landfill.

Based on the optimization results, the benefits of this integrated MSWM system are significant reduction of energy consumption and equivalent CO_2 emissions. The other benefits are increasing the share of renewable energy production and creating more jobs through construction of AD and incinerations units. However, this should be noted that in all these scenarios it has been assumed that different waste types are completely separated. Therefore, proper separation and sorting of recyclable material, food waste and yard waste is a big challenge. Another challenge is the lack of a specific policy for buying self-generated electricity, which reduces the interest from external investors to invest into the construction of AD and incineration units in the city. The other challenge is low public awareness about the dangers of landfilling and their important role in having an efficient MSWM system. Finally, the severe weather conditions during the long winters of MTL could affect the efficiency of AD and composting and these units would need further thermal energy to operate properly.

Author Contributions: Data curation, T.M.; Formal analysis, T.M. and S.R.; Investigation, T.M. and S.R.; Methodology, T.M. and S.R.; Project administration, T.M; Resources, T.M; Software, T.M and S.R.; Supervision, U.E.; Validation, S.R.; Visualization, T.M. and S.R.; Writing – original draft, T.M. and S.R.; Writing – review & editing, U.E. All authors have read and agreed to the final version of the manuscript.

Funding: This research was supported by NSERC funding for the Canada Excellence Research Chair in Smart, Sustainable and Resilient Cities and Communities.

Conflicts of Interest: The authors declare no conflict of interest.

References

1. Mayer, F.; Bhandari, R.; Gath, S. Critical review on life cycle assessment of conventional and innovative waste-to-energy technologies. *Sci. Total Environ.* **2019**, *672*, 708–721. [CrossRef] [PubMed]

2. Malmir, T.; Tojo, Y. Municipal solid waste management in Tehran: Changes during the last 5 years. *Waste Manag. Res.* **2016**, *34*, 449–456. [CrossRef] [PubMed]

3. Sindhu, R.; Gnansounou, E.; Rebello, S.; Binod, P.; Varjani, S.; Thakur, I.S.; Nair, R.B.; Pandey, A. Conversion of food and kitchen waste to value-added products. *J. Environ. Manag.* **2019**, *241*, 619–630. [CrossRef] [PubMed]

4. United States Environmental Protection Agency. *Advancing Sustainable Materials Management: 2014 Fact Sheet*; United States Environmental Protection Agency, Office of Land and Emergency Management: Washington, DC, USA, 2016.

5. United States Environmental Protection Agency. *Advancing Sustainable Materials Management: 2015 Fact Sheet*; United States Environmental Protection Agency, Office of Land and Emergency Management: Washington, DC, USA, 2018.

6. Evangelisti, S.; Lettieri, P.; Borello, D.; Clift, R. Life cycle assessment of energy from waste via anaerobic digestion: A UK case study. *Waste Manag.* **2014**, *34*, 226–237. [CrossRef]

7. European Environment Agency. *Diversion of Waste from Landfill*; European Environment Agency: Copenhagen, Denmark, 2019.

8. Levaggi, L.; Levaggi, R.; Marchiori, C.; Trecroci, C. Waste-to-Energy in the EU: The Effects of Plant Ownership, Waste Mobility, and Decentralization on Environmental Outcomes and Welfare. *Sustainability* **2020**, *12*, 5743. [CrossRef]

9. McDougall, F.R.; White, P.R.; Franke, M.; Hindle, P. *Integrated Solid Waste Management: A Life Cycle Inventory*; John Wiley & Sons: Hoboken, NJ, USA, 2001.

10. Lundie, S.; Peters, G.M. Life cycle assessment of food waste management options. *J. Clean. Prod.* **2005**, *13*, 275–286. [CrossRef]

11. De Feo, G.; Ferrara, C.; Iuliano, C.; Grosso, A. LCA of the Collection, Transportation, Treatment and Disposal of Source Separated Municipal Waste: A Southern Italy Case Study. *Sustainability* **2016**, *8*, 1084. [CrossRef]

12. Lyng, K.-A.; Stensgård, A.E.; Hanssen, O.J.; Modahl, I.S. Relation between greenhouse gas emissions and economic profit for different configurations of biogas value chains: A case study on different levels of sector integration. *J. Clean. Prod.* **2018**, *182*, 737–745. [CrossRef]

13. Nayal, F.S.; Mammadov, A.; Ciliz, N. Environmental assessment of energy generation from agricultural and farm waste through anaerobic digestion. *J. Environ. Manag.* **2016**, *184*, 389–399. [CrossRef]

14. Grigatti, M.; Barbanti, L.; Hassan, M.U.; Ciavatta, C. Fertilizing potential and CO_2 emissions following the utilization of fresh and composted food-waste anaerobic digestates. *Sci. Total Environ.* **2020**, *698*, 134198. [CrossRef]

15. Quek, A.; Balasubramanian, R. Life Cycle Assessment of Energy and Energy Carriers from Waste Matter—A Review. *J. Clean. Prod.* **2014**, *79*, 18–31. [CrossRef]

16. Gouvernement du Québec. *The 2030 Energy Policy*; Gouvernement du Québec: Montreal, QC, Canada, 2016.

17. Mohsenizadeh, M.; Tural, M.K.; Kentel, E. Municipal solid waste management with cost minimization and emission control objectives: A case study of Ankara. *Sustain. Cities Soc.* **2020**, *52*, 101807. [CrossRef]

18. Richter, A.; Bruce, N.; Ng, K.T.; Chowdhury, A.; Vu, H.L. Comparison between Canadian and Nova Scotian waste management and diversion models—A Canadian case study. *Sustain. Cities Soc.* **2017**, *30*, 139–149. [CrossRef]

19. Ville de Montréal. *Portrait 2016 des Matières Résiduelles de L'agglomération de Montreal*; Ville de Montréal: Montreal, QC, Canada, 2017.

20. Québec, Government of Québec, Institut de la statistique. 2015. Available online: www.stat.gouv.qc.ca/statistiques/profils/region_06/region_06_00_an.htm (accessed on 23 October 2019).

21. Communauté Métropolitaine de Montréal. *Draft amendment, To the 2015–2020 residual materials management plan for metropolitan Montréal*; Communauté Métropolitaine de Montréal: Montreal, QC, Canada, 2019.

22. Henault-Ethier, L.; Martin, J.P.; Housset, J. A dynamic model for organic waste management in Quebec (D-MOWIQ) as a tool to review environmental, societal and economic perspectives of a waste management policy. *Waste Manag* **2017**, *66*, 196–209. [CrossRef]

23. Ville de Montréal. *Bilan 2017 des Matières Résiduelles de L'agglomération de Montréal*; Ville de Montréal: Montreal, QC, Canada, 2018.

24. Nielsen, T.D.; Holmberg, K.; Stripple, J. Need a bag? A review of public policies on plastic carrier bags—Where, how and to what effect? *Waste Manag.* **2019**, *87*, 428–440. [CrossRef]

25. Malmir, S.; Montero, B.; Rico, M.; Barral, L.; Bouza, R. Morphology, thermal and barrier properties of biodegradable films of poly (3-hydroxybutyrate-co-3-hydroxyvalerate) containing cellulose nanocrystals. *Compos. Part A Appl. Sci. Manuf.* **2017**, *93*, 41–48. [CrossRef]

26. Malmir, S.; Montero, B.; Rico, M.; Barral, L.; Bouza, R.; Farrag, Y. PHBV/CNC bionanocomposites processed by extrusion: Structural characterization and properties. *Polym. Compos.* **2019**, *40*, E275–E284. [CrossRef]

27. Guerin, J.E.; Paré, M.C.; Lavoie, S.; Bourgeois, N. The importance of characterizing residual household waste at the local level: A case study of Saguenay, Quebec (Canada). *Waste Manag.* **2018**, *77*, 341–349. [CrossRef]

28. EPIC/CSR (Environment and Plastics Industry Council/Corporations Supporting Recycling). *Integrated Solid Waste Management Tools: User Guidance Document*; EPIC/CSR (Environment and Plastics Industry Council/Corporations Supporting Recycling): Carrey, NC, USA, 2000.

29. Haight, M. *Technical Report: Integrated Solid Waste Management Model*; University of Waterloo: Waterloo, ON, Canada, 2004.

30. Pourreza Movahed, Z.; Kabiri, M.; Ranjbar, S.; Joda, F. Multi-objective optimization of life cycle assessment of integrated waste management based on genetic algorithms: A case study of Tehran. *J. Clean. Prod.* **2020**, *247*, 119153. [CrossRef]

31. Éco Entreprises Québec. *Consultation Tarif 2019—Sommaire*; Éco Entreprises Québec: Montreal, QC, Canada, 2019.

32. Colvero, D.A.; Ramalho, J.; Gomes, A.P.D.; de Matos, M.A.A.; da Cruz Tarelho, L.A. Economic analysis of a shared municipal solid waste management facility in a metropolitan region. *Waste Manag.* **2020**, *102*, 823–837. [CrossRef]

33. Deb, K. *Multi-objective Optimization Using Evolutionary Algorithms*; John Wiley & Sons: Hoboken, NJ, USA, 2001; Volume 16.

34. Deb, K.; Pratap, A.; Agarwal, S.; Meyarivan, T.A.M.T. A Fast and Elitist Multiobjective Genetic Algorithm: NSGA-II. *IEEE Trans. Evol. Comput.* **2002**, *6*, 182–197. [CrossRef]

35. Wang, Y.; Shen, Y.; Zhang, X.; Cui, G.; Sun, J. An Improved Non-dominated Sorting Genetic Algorithm-II (INSGA-II) applied to the design of DNA codewords. *Math. Comput. Simul.* **2018**, *151*, 131–139. [CrossRef]

36. Thiriet, P.; Bioteau, T.; Tremier, A. Optimization method to construct micro-anaerobic digesters networks for decentralized biowaste treatment in urban and peri-urban areas. *J. Clean. Prod.* **2020**, *243*, 118478. [CrossRef]

37. Christensen, T.H.; Damgaard, A.; Levis, J.; Zhao, Y.; Björklund, A.; Arena, U.; Barlaz, M.A.; Starostina, V.; Boldrin, A.; Astrup, T.F.; et al. Application of LCA modelling in integrated waste management. *Waste Manag.* **2020**, *118*, 313–322. [CrossRef]

38. Habib, K.; Schmidt, J.H.; Christensen, P. A historical perspective of Global Warming Potential from Municipal Solid Waste Management. *Waste Manag.* **2013**, *33*, 1926–1933. [CrossRef] [PubMed]

39. Bartocci, P.; Zampilli, M.; Liberti, F.; Pistolesi, V.; Massoli, S.; Bidini, G.; Fantozzi, F. LCA analysis of food waste co-digestion. *Sci. Total Environ.* **2020**, *709*, 136187. [CrossRef]

40. Safdar, N.; Khalid, R.; Ahmed, W.; Imran, M. Reverse logistics network design of e-waste management under the triple bottom line approach. *J. Clean. Prod.* **2020**, *272*, 122662. [CrossRef]

41. Sooriyaarachchi, T.M.; Tsai, I.T.; El Khatib, S.; Farid, A.M.; Mezher, T. Job creation potentials and skill requirements in, PV, CSP, wind, water-to-energy and energy efficiency value chains. *Renew. Sustain. Energy Rev.* **2015**, *52*, 653–668. [CrossRef]

42. Walker, M.; Theaker, H.; Yaman, R.; Poggio, D.; Nimmo, W.; Bywater, A.; Blanch, G.; Pourkashanian, M. Assessment of micro-scale anaerobic digestion for management of urban organic waste: A case study in London, UK. *Waste Manag.* **2017**, *61*, 258–268. [CrossRef]

Publisher's Note: MDPI stays neutral with regard to jurisdictional claims in published maps and institutional affiliations.

Article

Evaluation of Anaerobic Digestion of Dairy Wastewater in an Innovative Multi-Section Horizontal Flow Reactor

Marcin Dębowski [1],*, Marcin Zieliński [1], Marta Kisielewska [1] and Joanna Kazimierowicz [2]

[1] Department of Environmental Engineering, Faculty of Geoengineering, University of Warmia and Mazury in Olsztyn, 10-719 Olsztyn, Poland; marcin.zielinski@uwm.edu.pl (M.Z.); jedrzejewska@uwm.edu.pl (M.K.)

[2] Department of Water Supply and Sewage Systems, Faculty of Civil Engineering and Environmental Sciences, Bialystok University of Technology, 15-351 Bialystok, Poland; j.kazimierowicz@pb.edu.pl

* Correspondence: marcin.debowski@uwm.edu.pl

Received: 11 April 2020; Accepted: 6 May 2020; Published: 11 May 2020

Abstract: The aim of this study was the performance evaluation of anaerobic digestion of dairy wastewater in a multi-section horizontal flow reactor (HFAR) equipped with microwave and ultrasonic generators to stimulate biochemical processes. The effects of increasing organic loading rate (OLR) ranging from 1.0 g chemical oxygen demand (COD)/L·d to 4.0 g COD/L·d on treatment performance, biogas production, and percentage of methane yield were determined. The highest organic compounds removals (about 85% as COD and total organic carbon—TOC) were obtained at OLR of 1.0–2.0 g COD/L·d. The highest biogas yield of 0.33 ± 0.03 L/g COD removed and methane content in biogas of 68.1 ± 5.8% were recorded at OLR of 1.0 g COD/L·d, while at OLR of 2.0 g COD/L·d it was 0.31 ± 0.02 L/COD removed and 66.3 ± 5.7%, respectively. Increasing of the OLR led to a reduction in biogas productivity as well as a decrease in methane content in biogas. The best technological effects were recorded in series with an operating mode of ultrasonic generators of 2 min work/28 min break. More intensive sonication reduced the efficiency of anaerobic digestion of dairy wastewater as well as biogas production. A low nutrient removal efficiency was observed in all tested series of the experiment, which ranged from 2.04 ± 0.38 to 4.59 ± 0.68% for phosphorus and from 9.67 ± 3.36 to 20.36 ± 0.32% for nitrogen. The effects obtained in the study (referring to the efficiency of wastewater treatment, biogas production, as well as to the results of economic analysis) proved that the HFAR can be competitive to existing industrial technologies for food wastewater treatment.

Keywords: dairy wastewater; biogas; anaerobic digestion; anaerobic horizontal flow reactor; microwave radiation; ultrasound

---

1. Introduction

Bioenergy production from waste substrates supports the circular economy [1] and contributes to improvement of economic and ecological indicators [2]. Anaerobic digestion (AD) is an effective method for biogas production from wastewater with high concentration organic compounds, e.g., from meat industry, the fruit and vegetable sector, the sugar industry, and dairy processing [3]. Currently, many different constructions and solutions of anaerobic bioreactors are used in wastewater treatment systems. The commonly used reactors types are UASB (Upflow Anaerobic Sludge Blanket) ones with biomass in the form of granules and EGSB (Expanded Granular Sludge Bed) ones as a variant of UASB reactor. Their disadvantages are the long start-up period (lasting from 3 to 5 months) and the necessity to pre-treat wastewater with high fat and suspensions concentrations [4]. Frequently used reactors are CSTRs (Continuous-flow Stirred Tank Reactors) with suspended biomass. The operational problems of CSTRs result from the flotation of biomass in settling tanks and its washing out from the reaction chamber. The low concentration of biomass determines the reactors' operation at relatively low OLRs [5]. Other reactors are anaerobic biological beds (ABBR—Anaerobic Biological Bed Reactors) characterized by long start-up period due to the prolonged formation of anaerobic biological membrane and large dimensions. However, a risk of clogging the filter spaces often takes place [6]. In turn, Anaerobic Fluidized Biological Bed (AFBB) reactors are very energy-consuming as a result of intensive wastewater recirculation [7]. In hybrid reactors, the advantages of individual anaerobic reactors are combined [8]. The problems and limitations of anaerobic reactors encourage a search for improvements to make them more effective and economic [9]. According to the literature, AD can be effectively intensified by physical factors, including microwave radiation (MR), ultrasound (US), constant magnetic field (CMF), as well as electromagnetic field (EM) [10].

Considering previous studies, MR and US have been demonstrated as the promising technologies enhancing AD [11]. MR can be used as an alternative heating method for anaerobic reactors [12]. Advantages of technologies based on MR include high heating effectiveness, fast heat transfer, selective and uniform heating performance, short reaction time, easy operation, increasing the speed of biochemical reactions, and low formation of hazardous products [13]. Moreover, MR produces non-thermal effects, which means a number of phenomena that appear in heated bodies, and whose presence or intensity cannot be explained only by heating alone. In AD, these phenomena are the acceleration of enzymatic reactions, as well as the formation of specific anaerobic bacterial communities [14]. Many literature data presents numerous technologies assisted by US [15]. Ultrasonic irradiation has been successfully used in ammonia recovery from wastewater [16], in increasing biodegradation of organic pollutants [17], in disinfecting water [18], in supporting the membrane filtration process [19] and in preparing sewage sludge before fermentation or dewatering [20]. US generates monolithic cavitation by passage of ultrasonic waves through the liquid medium, which results in various physical and chemical changes in liquid solutions [21]. Thus, US can be used for degassing of anaerobic sludge and removing the gaseous products from the reaction chamber. Using of the small and carefully selected doses of ultrasonic energy may allow to break down the compact structure of anaerobic flocs and granules, which finally enhance the biogas discharge from anaerobic reactors.

Ultrasound application in wastewater treatment technology aiming at the intensification of biogas and methane production has been confirmed in preliminary research as well as by other researchers [22,23]. However, recent studies have not specified clearly the parameters of sonication process for achieving the highest efficiency of wastewater treatment and biogas production with high methane content. Moreover, laboratory tests often do not provide reliable data to design and operate installations on an industrial scale. Thus, there is a need to study a pilot-scale or semi-technical scale installations.

The purpose of the study was to determine the efficiency of dairy wastewater treatment and the biogas production in the multi-section horizontal flow anaerobic reactor (HFAR) equipped with MR and US devices.

2. Materials and Methods

2.1. Research Station

The experiments were carried out in the multi-section horizontal flow anaerobic reactor (HFAR), (Figure 1). The internal axis of reactor was a shaft rotating at 1 rpm, to which the stirring rods were attached. The reaction chamber was divided with perforated partitions into the four sections.

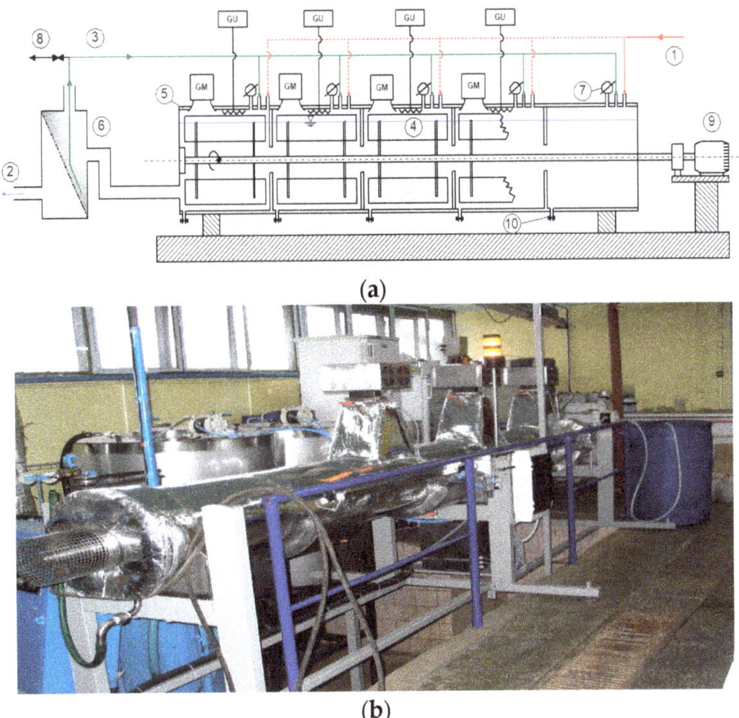

(a)

(b)

Figure 1. Horizontal flow anaerobic reactor (HFAR) (**a**) schematic diagram, (**b**) photo. 1—influent; 2—effluent; 3—sludge recirculation; 4—mixing blade; 5—thermal insulation; 6—membrane module; 7—gas meter; 8—digestate; 9—gearmotor; 10—sampling port; GM—microwave generator; GU—ultrasonic generator.

The HFAR design parameters were as follows: total length—4.0 m, total diameter—0.3 m, total cross-sectional area—0.070 m^2, working cross-sectional area—0.063 m^2, height of filling—0.25 m, total volume—0.28 m^3, working volume—0.25 m^3.

The reactor was heated using microwave generators (GM), (Plasmatronics) with magnetrons (Figure 2a). There was a possibility of smooth regulation from 0 to 800 W. This type of magnetrons is commonly used in microwave heating techniques due to their high performance, low price, and small dimensions. Electricity was converted into microwave energy with a conversion efficiency of 52% at 2.45 GHz. The positive effects of microwaves on biological processes with using magnetrons has also been confirmed in other studies [24]. A microprocessor integrated with temperature detectors controlled the temperature inside each section of the HFAR. The temperature was maintained at 38 °C, and when it dropped below, the microwave generators automatically started. Their work was finished when the defined temperature inside the reactor was achieved.

The ultrasonic generators (GU) used in the study were installed at each section of the HFAR (Figure 2b). Sections were equipped with a package of the four ultrasonic transducers with a frequency of 20 kHz and a power of 95 W. Each GU was integrated with a control panel to control its work. In the upper part of HFAR, in all sections, there were ports for substrate supplementation, sludge, and biogas collection, as well as the temperature and pH measurement, which created the possibility of wastewater dosing to the various sections of HFAR, anaerobic sludge recirculation, and biogas analyzing. The biogas was collected in the sealed containers. The dairy wastewater was dosed by pumps to the first section of reactor. The digestate was collected in the thickener with working volume of 3.0 m^3. The sludge from the bottom of the thickener was recycled and then mixed with the raw wastewater. Effluent was discharged outside the technological system.

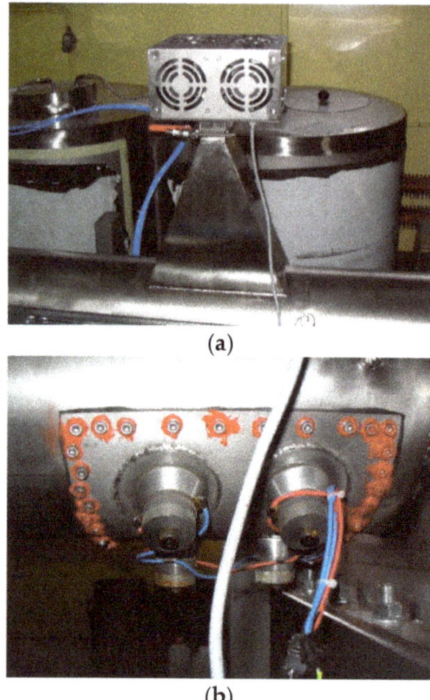

(a)

(b)

Figure 2. Reactor equipment (**a**) microwave generator, (**b**) ultrasound generator.

2.2. Concept of Experimental Design

The study was divided into four stages differing in OLR applied: stage 1—OLR of 1.0 g COD/L d, stage 2—2.0 g COD/L d, stage 3—3.0 g COD/L d, stage 4—4.0 g COD/L d. Each stage of the experiment was additionally divided into the four series differing in the sonication time regimes. In series 1, AD was carried out without the use of ultrasounds. In series 2, the operating mode of ultrasonic generators was 10 min work/20 min break, while in series 3 and 4, it was respectively 5 min work/25 min break and 2 min work/28 min break. A constant processing temperature of 38 ± 1 °C was maintained. The hydraulic retention time (HRT) was 24 h. Each series of the experiment lasted 30 days, which allowed for 30-fold exchange of the reactors content.

2.3. Materials

The wastewater used for AD originated from the retention tank of a dairy processing plant and was diluted with tap water to achieve the required OLRs. The characteristics of wastewater used in each stage of the experiment is shown in Table 1.

Anaerobic sludge from an anaerobic reactor treating the dairy wastewater and exploited on a technical scale was used as inoculum in the study. The concentration of inoculum in HFAR was 44 g total solids (TS)/L. The characteristics of the anaerobic inoculum used in the study is presented in Table 2.

Table 1. Characteristics of dairy wastewater used in the experiment.

Stage	Parameter				
	COD (mg/L)	TOC (mg/L)	TP (mg/L)	TN (mg/L)	pH
Raw Wastewater	5138.4 ± 53.8	1780.9 ± 17.3	190.1 ± 7.2	164.3 ± 6.7	9.06 ± 0.24
1	1029 ± 20.5	347.3 ± 15.7	46.2 ± 4.7	39.1 ± 4.5	7.13 ± 0.25
2	2014 ± 40.4	703.2 ± 18.2	79.3 ± 5.8	65.3 ± 5.7	7.19 ± 0.22
3	3092 ± 50.2	1027.1 ± 20.1	112.4 ± 6.7	107.5 ± 6.4	7.09 ± 0.24
4	4046 ± 60.4	1402.3 ± 22.5	149.7 ± 7.9	129.4 ± 7.8	7.13 ± 0.21

Table 2. Characteristics of the anaerobic inoculum used in the study.

Parameter	Unit	Mean
Hydration	(%)	95.6 ± 1.3
Specific resistance to filtration	(m/kg)	$1.768 \times 10^{15} \pm 1.631 \times 10^{14}$
Capillary suction time	(s)	1072 ± 175
Total solids	(g/L)	44.3 ± 4.9
Mineral solids	(g/L)	19.0 ± 1.3
Volatile solids	(g/L)	25.3 ± 3.6
Filtrate COD	(mg/L)	624.9 ± 84.0
Orthophosphates in filtrate	(mg P-PO$_4$/L)	105.6 ± 24.0
TN in filtrate	(mg TN/L)	163.8 ± 17.5
AN in filtrate	(mg N-NH$_4$/L)	144.2 ± 25.3
pH	-	7.75 ± 0.16

2.4. Analytical Methods and Statistical Procedures

The influent and effluent were analyzed once a day for COD, total phosphorus (TP), total nitrogen (TN) using the cuvette tests for spectrophotometer DR 2800 with mineralizer (HACH Lange, Düsseldorf, Germany), VSS according to the gravimetric method (part E of EPA Standard Method 2540), TOC with the use of TOC 1200 analyzer (Thermo Scientific, Waltham, USA). The analyzer determined the content of TC in the sample by combustion and then in the same sample determined the IC content in infrared analytical cycle. The organic carbon content in the sample was calculated by subtracting the TC value and IC value. The pH was determined by using the pH-meter VWR 1000 L.

The biomass yield was determined as the ratio of the amount of biomass produced to the amount of COD removed. The biomass production was determined as the change in biomass concentration in the reactor and the biomass concentration in the effluent. In order to determine the concentration of biomass in the whole reactor, samples were taken from each reactor section and then the measured concentrations were averaged. The biomass yield was estimated as follows (1):

$$Y_{Pb} = \frac{VSS' - VSS}{C - C_e} \tag{1}$$

where: VSS'—sum of the concentration of VSS in the reactor and VSS in the effluent in the next day (g/L); VSS—sum of the concentration of VSS in the reactor and VSS in the effluent in the current day (g/L); C—the concentration of COD in the influent (g/L); C_e—the concentration of COD in the effluent (g/L).

Throughout the experiment, the amount of biogas was monitored by biogas meters installed at each reactor's section. The composition of biogas produced at each section of reactor was measured every 24 h using a gastight syringe (20 mL injection volume) and a gas chromatograph (GC, 7890A Agilent) equipped with a thermal conductivity detector (TCD). The GC was fitted with the two Hayesep Q columns (80/100 mesh), two molecular sieve columns (60/80 mesh), and Porapak Q column (80/100) operating at a temperature of 70 °C. The temperature of the injection and detector ports were 150 °C and 250 °C, respectively. Helium and argon were used as the carrier gases at a flow of 15 mL/min.

Additionally, biogas was analyzed by the GMF 430 Gas Data analyzer. The content of methane (CH_4), and carbon dioxide (CO_2) were measured.

The results were processed statistically with the Statictica 13.1 PL package (StatSoft, Inc., Tulsa, OK, USA). The hypothesis on the distribution of each analyzed variable was verified based on the W Shapiro–Wilk's test. One way analysis of variance (ANOVA) was conducted to determine the significance of differences between the variables. The homogeneity of variance in groups was tested with Levene's test, whereas Tukey's RIR test was used to determine the significance of differences between the analyzed variables. In all tests, differences were considered significant at $p = 0.05$.

3. Results and Discussion

3.1. Organic Compounds and Nutrient Removal

The application of increasing OLRs to anaerobic reactor directly influences the reduction of its performance. In the study, the highest COD and TOC removal efficiencies higher than 85% were found at OLR of 2.0 g COD/L·d (Table 3). Comparable effects were obtained at OLR of 1.0 g COD/L·d. Further increase in OLR caused a deterioration in wastewater treatment efficiency. The lowest removal efficiencies of COD and TOC were recorded in stage 4 at OLR of 4.0 g COD/L·d (Table 3).

Table 3. The efficiency of organic compounds and nutrient removal from dairy wastewater.

Stage	Series	Removal Efficiency (%)				Load Removal (g/d)			
		COD	TOC	TN	TP	COD	TOC	TN	TP
1	1	73.26 ± 1.63	69.51 ± 1.77	13.46 ± 3.20	2.84 ± 0.33	183.15 ± 18.25	60.35 ± 7.13	1.32 ± 0.11	0.33 ± 0.05
	2	65.78 ± 1.71	64.38 ± 1.35	9.67 ± 3.36	2.04 ± 0.38	164.45 ± 14.54	55.90 ± 6.41	0.95 ± 0.12	0.24 ± 0.06
	3	80.93 ± 1.26	78.66 ± 1.42	13.88 ± 4.07	2.93 ± 0.27	202.33 ± 21.22	68.30 ± 6.62	1.36 ± 0.23	0.34 ± 0.05
	4	84.64 ± 1.73	84.94 ± 0.56	14.52 ± 2.86	3.06 ± 0.11	211.60 ± 20.05	73.75 ± 6.81	1.42 ± 0.24	0.35 ± 0.07
2	1	78.18 ± 1.09	80.22 ± 1.51	18.69 ± 0.61	3.72 ± 0.37	390.88 ± 25.11	141.03 ± 8.91	3.05 ± 0.61	0.74 ± 0.12
	2	69.26 ± 1.52	76.00 ± 1.22	13.46 ± 0.61	2.68 ± 0.18	346.28 ± 23.24	133.60 ± 8.44	2.20 ± 0.53	0.53 ± 0.08
	3	82.15 ± 0.92	83.18 ± 1.29	18.42 ± 0.07	3.67 ± 0.13	410.73 ± 25.13	146.23 ± 9.02	3.01 ± 0.60	0.73 ± 0.13
	4	85.13 ± 1.05	85.62 ± 1.04	20.36 ± 0.62	4.05 ± 0.04	425.65 ± 24.21	150.53 ± 9.92	3.32 ± 0.63	0.80 ± 0.15
3	1	66.93 ± 0.80	68.69 ± 0.76	16.44 ± 0.40	3.86 ± 1.14	501.95 ± 21.81	176.38 ± 10.11	4.42 ± 0.70	1.08 ± 0.21
	2	61.94 ± 0.85	61.33 ± 0.67	12.68 ± 0.01	2.98 ± 1.01	464.55 ± 16.44	157.48 ± 10.21	3.41 ± 0.62	0.84 ± 0.18
	3	75.73 ± 0.84	77.33 ± 0.76	16.53 ± 0.14	3.88 ± 0.97	567.98 ± 19.23	198.58 ± 11.04	4.44 ± 0.71	1.09 ± 0.22
	4	84.28 ± 0.74	84.70 ± 0.82	19.55 ± 0.66	4.59 ± 0.68	632.13 ± 24.31	217.50 ± 12.03	5.25 ± 0.82	1.29 ± 0.25
4	1	63.84 ± 0.55	66.31 ± 0.56	17.92 ± 0.22	3.57 ± 0.81	638.43 ± 26.42	232.48 ± 12.64	5.80 ± 0.80	1.33 ± 0.31
	2	52.75 ± 0.42	56.98 ± 0.48	11.69 ± 0.28	2.33 ± 0.75	527.45 ± 25.24	199.75 ± 10.61	3.78 ± 0.63	0.87 ± 0.18
	3	68.15 ± 0.53	72.70 ± 0.33	18.12 ± 0.01	3.61 ± 0.75	681.53 ± 23.64	254.88 ± 13.92	5.86 ± 0.81	1.35 ± 0.40
	4	77.19 ± 0.43	78.17 ± 0.36	19.38 ± 0.15	3.86 ± 0.51	771.90 ± 23.53	274.05 ± 14.63	6.27 ± 0.73	1.44 ± 0.35

In the study, the most effective operating mode of GU was 2 min work/28 min break. Generally, the application of GU enhanced COD and TOC removal in all experimental stages in series 3 and 4. However, the highest efficiencies of removal were noted at OLR ranging from 1 g COD/L·d to 2 g COD/L·d (Table 3, Figure 3). The operating mode of the ultrasonic generators of 10 min work/20 min break was too long to enhance biological processes. Thus, in series 2 of all stages of the experiment, the efficiencies of organic compounds removal were lower than that in the control sample (Table 3, Figure 3).

Regardless of the experimental series, low efficiencies of nitrogen and phosphorus removal were obtained (Table 3). The highest efficiency of TN removal was found in series 4 of stage 2 and it was 20.36 ± 0.62% (52.0 ± 4.1 mg/L in the effluent), while the lowest of 9.67 ± 3.36% was obtained in series 2 of the stage 2 (56.5 ± 4.5 mg/L in the effluent) (Table 3, Figure 3). However, the highest efficiency of TP removal of 4.59 ± 0.68% (107.2 ± 7.2 mg/L in the effluent) was recorded in series 4 of stage 3. The lowest efficiency of TP removal was noted in series 2 of stage 2, and it was as low as 2.04 ± 0.38% (77.2 ± 5.8 mg/L in the effluent) (Table 3, Figure 3). Other authors also confirm low nutrient removal in anaerobic conditions, which is limited to the nutrient demand for biomass growth [25,26].

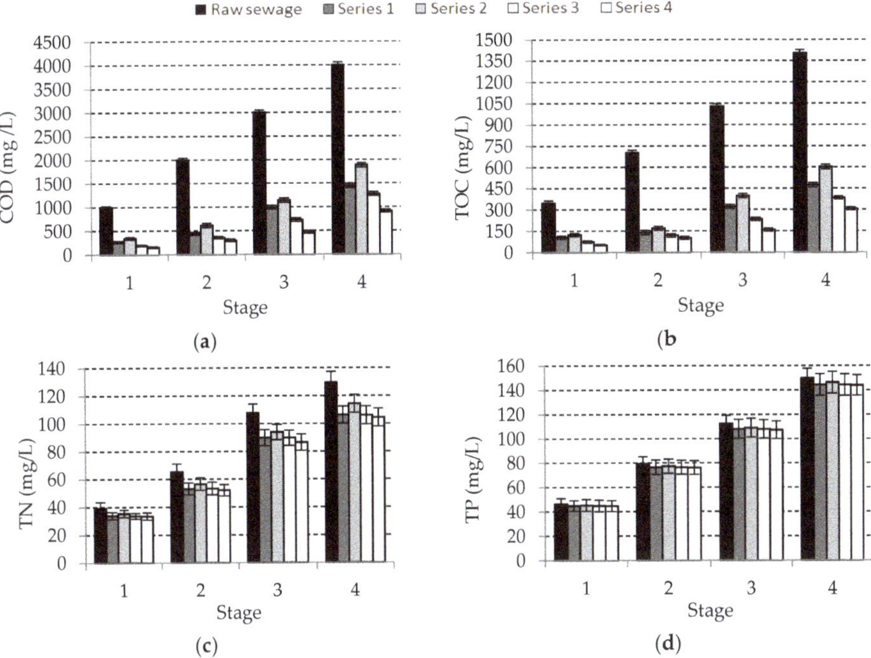

Figure 3. Concentrations of analyzed parameters in the effluent (**a**) COD, (**b**) TOC, (**c**) TN and (**d**) TP.

The effectiveness of dairy wastewater treatment in anaerobic conditions was also analyzed by other researchers [27]. After two years of experiments, they obtained a reduction of about 90% in COD content, exploiting an anaerobic filter at OLR of 5–6 g COD/L·d. Moreover, the effect of the HRT from 20 d at the beginning of the study to 3.33 d at the end was also analyzed. Shortening the HRT did not decrease the efficiency of organic compounds' removal, indicating a steady-state of reactor. Other authors also found high efficiency of organic compounds removal, up to 98% from whey using anaerobic filters [28]. At HRT of 142 days and OLR of 9.8 g COD/L·d, high treatment efficiency in mesophilic conditions was obtained. While testing a wide range of OLRs from 7.9 to 45.42 g COD/L·d, the removal efficiency of organic compounds from whey was as high as 98% [29].

Anaerobic reactors are often used for treating wastewater from dairy processing because of its high content of easily degradable organic compounds [30]. However, the efficiency of anaerobic wastewater treatment is a result of many factors, including reactor design, process temperature, and wastewater composition. For example, the treatment of textile wastewater by using an UASB reactor allowed to achieve COD removal efficiency of 9% to 51% [31]. The treatment of starch wastewater resulted in a reduction of COD in the range of 77% to 93% [32]. Other studies examined the effects of OLRs (0.82 and 6.11 g COD/L·d) and HRT ranging from 4.1 d to 1.7 d on UASB reactor performance using wastewater with a COD concentration of 10 g/L [33]. In mesophilic conditions, the COD removal efficiency varied from 90% to 97%. The obtained results are similar to those with anaerobic trickling filters. At OLR of 4.45 g COD/L·d and HRT of 2.64 d, removal efficiency of COD was 81% [34]. The OLR during anaerobic degradation of acid whey ranged from 1.6 to 12.8 g COD/L·d. The efficiency depended on the OLR. The highest COD removal rate was observed at the lowest OLR applied (about 100%) and at maximum OLR, COD removal efficiency was 68% [35].

3.2. Biogas Yield and Its Composition

The lowest biogas production was observed in stage 1 at OLR of 1.0 g COD/L·d, ranging from 48.18 ± 4.5 L/d in series 2 to 70.25 ± 6.1 L/d in series 4 (Figure 4). The biogas yield in these series was from 0.29 ± 0.02 L/g COD removed and 0.86 ± 0.11 L/g TOC removed to 0.33 ± 0.03 L/g COD removed and 0.95 ± 0.12 L/g TOC removed (Table 4). Taking into account the amount of biogas produced per gram of COD or TOC removed, this stage of experiment was the most advantageous. In this stage, the highest methane content in biogas ranged from 63.4 ± 5.2% in series 1 to 68.1 ± 5.8% in series 4 was achieved (Figure 4).

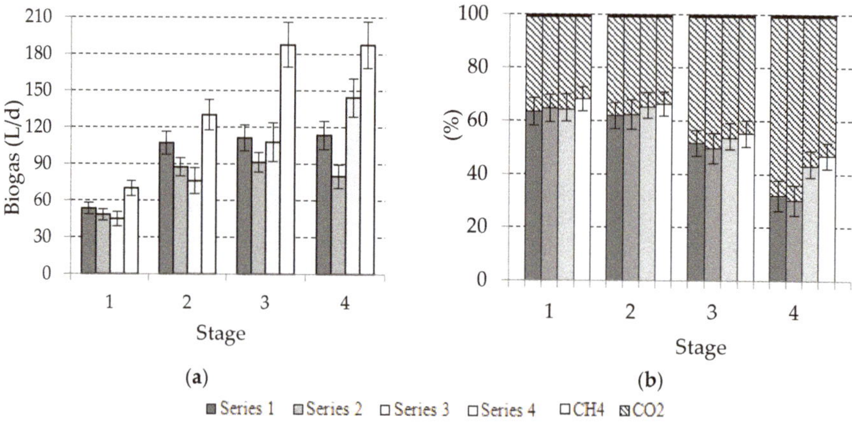

Figure 4. (**a**) Daily biogas production and (**b**) biogas composition depending on experimental stages and series.

Table 4. Biogas yield obtained in experimental series.

Stage	Series	Biogas Yield			
		L/g COD Added	L/g COD Removed	L/g TOC Added	L/g TOC Removed
1	1	0.21 ± 0.01	0.29 ± 0.02	0.61 ± 0.07	0.88 ± 0.12
	2	0.19 ± 0.01	0.29 ± 0.02	0.55 ± 0.08	0.86 ± 0.11
	3	0.26 ± 0.01	0.32 ± 0.02	0.74 ± 0.06	0.94 ± 0.13
	4	0.28 ± 0.02	0.33 ± 0.03	0.81 ± 0.09	0.95 ± 0.12
2	1	0.21 ± 0.01	0.27 ± 0.03	0.61 ± 0.04	0.76 ± 0.11
	2	0.18 ± 0.01	0.25 ± 0.02	0.50 ± 0.05	0.66 ± 0.10
	3	0.24 ± 0.01	0.29 ± 0.02	0.68 ± 0.07	0.82 ± 0.11
	4	0.26 ± 0.02	0.31 ± 0.02	0.74 ± 0.11	0.87 ± 0.12
3	1	0.15 ± 0.01	0.22 ± 0.01	0.43 ± 0.06	0.63 ± 0.09
	2	0.12 ± 0.01	0.20 ± 0.01	0.36 ± 0.04	0.58 ± 0.08
	3	0.20 ± 0.01	0.27 ± 0.01	0.59 ± 0.05	0.76 ± 0.08
	4	0.25 ± 0.02	0.30 ± 0.02	0.73 ± 0.07	0.86 ± 0.07
4	1	0.11 ± 0.01	0.18 ± 0.01	0.32 ± 0.04	0.49 ± 0.06
	2	0.08 ± 0.01	0.15 ± 0.01	0.23 ± 0.03	0.40 ± 0.05
	3	0.15 ± 0.01	0.22 ± 0.01	0.43 ± 0.04	0.59 ± 0.06
	4	0.19 ± 0.01	0.24 ± 0.01	0.54 ± 0.04	0.68 ± 0.08

In stage 2, the lowest biogas production was obtained in series 2 and it was 87.61 ± 7.4 L/d (0.25 ± 0.02 L/g COD removed, 0.66 ± 0.10 L/g TOC removed), and methane content in biogas reached 62.3 ± 5.5% (Figure 4, Table 4). In turn, the highest biogas production of 130.25 ± 12.5 L/d and the methane content of 66.3 ± 5.7% were noted in series 4 (Figure 4). The biogas yield was 0.31 ± 0.02 L/g COD removed and 0.87 ± 0.12 L/g TOC removed (Table 4). In stage 3, the lowest biogas production of 91.52 ± 7.9 L/d (0.20 ± 0.01 L/g COD removed, 0.58 ± 0.08 L/g TOC removed) and methane content of 49.7 ± 5.5% were obtained in series 2 (Figure 4, Table 4). In turn, the highest biogas production (187.74 ± 18.4 L/d, 0.30 ± 0.02 L/g COD removed, 0.86 ± 0.07 L/g TOC removed) and methane content (55.2 ± 5.8%) were noted in series 4 (Figure 4, Table 4). In stage 4, the lowest and the highest biogas production was as 80.17 ± 9.5 L/d (0.15 ± 0.01 L/g COD removed, 0.40 ± 0.05 L/g TOC removed) and 187.57 ± 19.1 L/d (0.24 ± 0.01 L/g COD removed, 0.68 ± 0.08 L/g TOC removed), respectively, while the methane content in biogas ranged from 30.2 ± 4.5% to 46.9 ± 4.8% (Figure 5, Table 4).

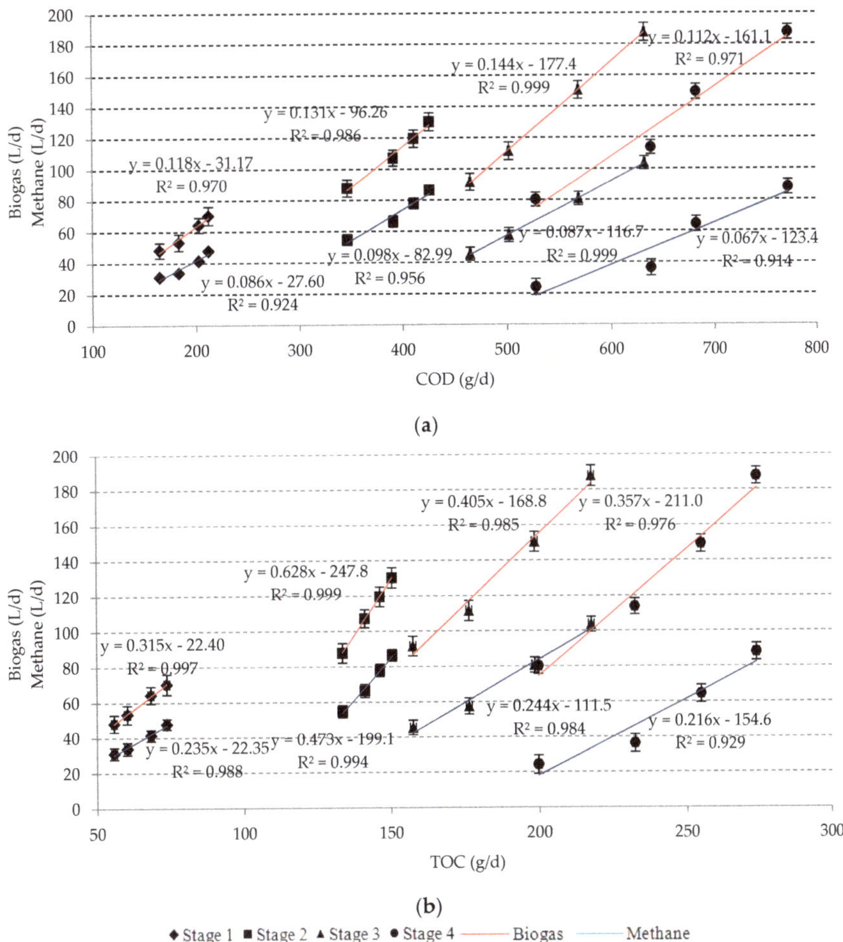

Figure 5. The relationship between biogas and methane production and organic compounds' removal (**a**) as COD, (**b**) as TOC.

It has been proven that sudden changes in temperatures significantly affect methanogenesis [36]. In this study, microwave heating was used as a method to create a stable thermal condition for anaerobic reactors [37]. While analyzing the effects of MR on biological processes, a positive impact on methanogenesis process was indicated [38]. Banik et al. (2003) studied the exposition of Methanosarcina barkeri DS-804 to microwave radiation of frequencies ranging from 13.5 to 36.5 for 2 h. Then the samples were incubated for 20 days. Radiated bacterial culture showed an increase in colony numbers and cell size. The biogas composition varied depending on the frequency of the radiation used. Maximum methane concentration in biogas was 76.5% at 31.5 GHz. In comparison to unirradiated setup, it was only 52.3%. Growth curves for both irradiated and unirradiated cultures were typical for the initial phase, the growth phase, as well as the stationary and decline phases. However, in irradiated culture, the growth phase was much faster. The results indicate that microwave radiation may induce specific metabolic activity of microorganisms to enhance their growth rate [38].

Parker et al. (1996) studied the effects of microwaves on the activity of a hydrated lipase enzyme. The enzyme was suspended in a solvent that was also the reactant. Then, the solution was heated to 50 °C by microwaves (2.45 GHz) and, for comparisation, conventionally. It was found that the rate of enzymatic reaction of irradiated enzyme increased up to 2–3 times in relation to conventional heating. However, other enzymes, such as lymphocytic protein kinases or phosphatases were inhibited by microwaves [39]. Other researchers found the increase of stability of immobilized enzymes in the microwave field [40,41]. According to the literature, the exposure of biomass to microwave radiation increased its solubilization and disrupted the network of exopolysaccharides, thus enhancing its subsequent anaerobic digestion [42]. Microwave pretreatment of waste activated sludge at frequency of 2.45 GHz increased the release of extracellular and intracellular biopolymers from the cells [43]. As a result, microwaved sludge produced more biogas in anaerobic digestion than conventionally pretreated sludge.

In the study, it was found that the main factor affecting biogas yield and composition was OLR. However, ultrasonication was also an important factor. There was a very strong relationship between biogas and methane production and the removal of organic compounds from dairy wastewater (Figure 5). Other authors [22] investigated the possibility of ultrasonic pretreatment for improving biogas production from olive mill wastewater. They found that using a low-frequency US enhanced both biogas production by about 20% and methane content in biogas. In other studies [23], US was used to support methane production from raw molasses wastewater; 441.6 L CH4/kg VS was obtained. The results demonstrated that US may be a suitable solution for treating this kind of wastewater. Low-frequency US pre-treatment significantly improved (by 40%) biogas and methane production from landfill leachate in anaerobic batch reactors [44]. In turn, other authors found that US enhanced biogas and methane production from the same wastewater only by 7% and 4%, respectively [45]. According to the literature, the application of ultrasound pretreatment had a positive effect on biogas yield from confectionery wastewater [46].

Ultrasounds have positive effects on biological processes and, on the other hand, they are often used to disrupt and inhibit microorganisms (e.g., in sludge stabilization and hygienization) [47,48]. An important issue is therefore to point appropriate ultrasound performance parameters. In the study, in series 2, the sonication process was in mode of 10 min work/20 min break, while in series 4, it was 2 min work/28 min break. The results revealed that 10 min of sonication was too long to enhance biological processes. The negative effects were probably related to partial cell destruction. Disintegration of the cells in various bacterial communities was also confirmed by other researchers [45]. The highest enhancement in biogas and methane production was reported in sonication time regime of 5 min work/25 min break and 2 min work/25 min break (Figure 5, Table 4). Ultrasounds improved the removal of gaseous products outside the reaction medium, and did not have a destructive effect on anaerobic microflora.

The results also confirmed the significant impact of the OLR value on the yields of biogas and methane production. Increase in OLR decreased biogas production per gram of COD and TOC removed and also influenced the reduction of methane content in biogas. This can be explained by volatile fatty acids (VFAs) accumulation in the reaction medium at high OLRs [49]. High VFAs concentrations caused a decrease in pH, resulting in inhibition of methanogens. Thus, the reduction in biogas and methane production were observed [50].

3.3. Biomass Growth

In the study, the highest biomass yield was recorded in stages 3 and 4, and it varied from 0.123 ± 0.01 g/g COD removed in series 2 of both stages to 0.1558 ± 0.01 g/g COD removed in series 1 of stage 4 and 0.1476 ± 0.01 g/g COD removed series 1 of stage 3 (Figure 6). A smaller increase in biomass growth yield was found at lower OLR values in stages 1 and 2. In stage 1, the biomass yield coefficient ranged from 0.0984 ± 0.01 g/g COD removed in series 2 to 0.1230 ± 0.01 g/g COD removed in series 1 of stage 1 (Figure 6). In stage 2, these values were similar and ranged from 0.1066 ± 0.01 g/g COD removed to 0.1312 ± 0.01 g/g COD removed in the same series (Figure 6). During the study, there were no significant differences in the biomass growth yield coefficients observed in the individual stages and series of experiment. Even in series 2, in which the biomass yields were the lowest, the differences were not significant (Figure 6).

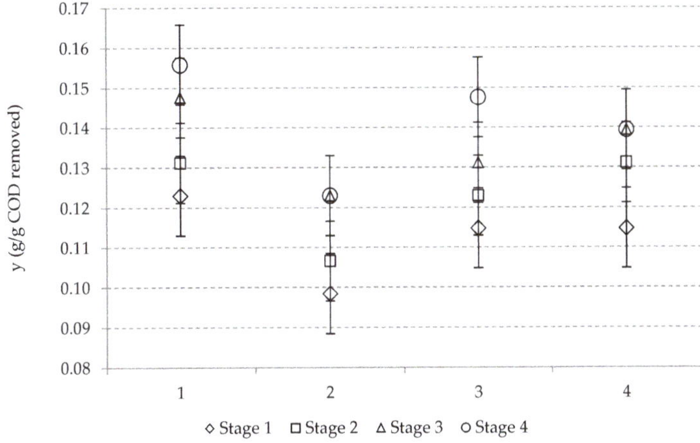

Figure 6. Biomass growth yield coefficients (y) experimental stages.

Anaerobic digestion is characterized by low biomass production (less excess sludge), which is 3 to 20 times lower than in aerobic treatment systems. Under anaerobic conditions, the yield of biomass per gram of COD removed is from 0.02 to 0.15 g/g, whereas in aerobic conditions, it was from 0.5 to 0.7 g/g [51].

3.4. Energy Balance

The energy balance of experimental series has been made to determine the amount of energy consumed for biogas production and for organic compound (as COD) removal. It can also be used to assess the whole technological and economic effectiveness of the presented technology. The reactor presented in the study corresponds to TRL 5 (Technology Readiness Level), which means that the technology was validated in a relevant environment. The basic equipment is integrated with the real supporting elements and the technology has been tested under simulated operating conditions. Thus, the estimated energy consumption can still be significantly optimized. The basic energy balance of wastewater treatment in HFAR was presented in Table 5.

Table 5. Energy balance of wastewater treatment in HFAR.

Stage	Series	Ultrasounds (kWh/d)	Mixing (kWh/d)	Microwave Heating (kWh/d)	Total Energy Demand (kWh/d)	Daily Biogas Production (L/d)	CH4 Concentration (%)	Daily CH4 Production (L/d)	Energetic Value of CH4 (kWh/L)	Total Energy Production (kWh/d)	Energy Balance (kWh/d)	Load of COD Removed (kg/d)	kwh/kg COD Removed
1	1	0	1.4	1.15	2.55	53.30	63.4	33.8		0.310	−2.240	0.183	12.230
	2	0.38			2.93	48.18	64.6	31.1		0.285	−2.645	0.164	16.084
	3	0.19			2.74	64.34	64.3	41.4		0.379	−2.361	0.202	11.669
	4	0.076			2.626	70.25	68.1	47.8		0.439	−2.187	0.211	10.336
2	1	0	1.4	1.15	2.55	107.10	61.9	66.3		0.608	−1.942	0.391	4.968
	2	0.38			2.93	87.61	62.3	54.6		0.500	−2.430	0.346	7.017
	3	0.19			2.74	119.52	65.1	77.8		0.714	−2.026	0.411	4.933
	4	0.076			2.626	130.25	66.3	86.4	0.00917	0.792	−1.834	0.423	4.309
3	1	0	1.4	1.15	2.55	111.43	51.6	57.5		0.527	−2.023	0.502	4.030
	2	0.38			2.93	91.52	49.7	45.5		0.417	−2.513	0.464	5.410
	3	0.19			2.74	150.51	53.6	80.7		0.740	−2.000	0.568	3.521
	4	0.076			2.626	187.74	55.2	103.6		0.950	−1.676	0.632	2.651
4	1	0	1.4	1.15	2.55	113.64	31.9	36.3		0.332	−2.218	0.638	3.474
	2	0.38			2.93	80.17	30.2	24.2		0.222	−2.708	0.527	5.134
	3	0.19			2.74	149.25	43.1	64.3		0.590	−2.150	0.681	3.155
	4	0.076			2.626	187.57	46.9	88.0		0.807	−1.819	0.772	2.357

The most favourable energy balance was noted in series 4, stage 3 (2 min sonification/28 min break. ORL 3.0 g COD/L·d). In this variant of the experiment, daily energy demand was 1.676 kWh/d (Table 5). The most energy-efficient variant without sonication was in series 1, stage 2, but energy consumption was higher (1.942 kWh/day). The least energy-efficient variant (−2.708 kWh/day) was at ORL of 4.0 g COD/L·d and the sonication regime of 10 min sonification/20 min break (Table 5). It was found that a unit energy consumption per unit of COD removed decreased, while OLR was increasing. In all stages of the experiment, the lowest unit energy expenditure requirement was observed in series 4 and ranged from 2.357 kWh/kg COD removed (ORL of 4.0 g COD/L·d) to 10.336 kWh/kg COD removed (ORL of 1.0 g COD/L·d).

In conclusion, the presented technology is competitive to other wastewater treatment solutions. Ozay et al. (2018) recorded a unit energy consumption of 7.80 kWh/kg COD removed [52]. Energy expenditure in the process of electrochemical wastewater treatment from the leather industry was 9.88 kWh/kg COD removed [53]. In studies on degradation of pyrrole in wastewater, the energy demand was 7.7 kWh/kg COD removed [54]. Ribera-Pi et al. (2018) operated a single-stage anaerobic membrane bioreactor (AnMBR) for dairy wastewater treatment and the unit energy consumption (kWh) per unit of COD removed (kg) was 2.4 [55]. Masłoń (2017) determined energy consumption for wastewater treatment plants ranging from 0.528 kWh/kg COD removed for plants with Imhoff tanks to 2.297 kWh/kg COD removed in SBR-type plants [56]. Generally, electric energy consumption in facilities below 2000 RLM was as 3.01 kWh/kg COD removed, while with size of RLM = 2000 ÷ 10.000, it was lower by half (1.54 kWh/kg COD removed) [57].

4. Conclusions

It was found that OLRs had the greatest impact on the effects of anaerobic digestion of dairy wastewater in terms of organic compound removal and biogas and methane yields. The highest efficiency of COD and TOC removal of 85.13 ± 1.05% and 85.62 ± 1.04%, respectively, was found at OLR of 2.0 g COD/L·d. The lowest organic compounds removals of 52.75 ± 0.42% as COD and 52.75 ± 0.42% as TOC were noted at OLR of 4.0 g COD/L·d.

The highest biogas yield of 0.33 ± 0.03 L/g COD removed and 0.95 ± 0.12 L/g TOC removed, and the methane content in biogas of 68.1 ± 5.8% were noted at OLR of 1.0 g COD/L·d. Comparable biogas production of 0.31 ± 0.02 L/g COD removed and 0.87 ± 0.12 L/g TOC removed (66.3 ± 5.7% of methane in biogas) was achieved at OLR of 2.0 g COD/L·d. Increase in OLR reduced biogas productivity and methane content in biogas. The lowest methane content was noted at OLR of 4.0 g COD/L·d and it ranged from 30.2 ± 4.5% to 46.9 ± 4.8% depending on the experimental series.

The study showed a significant impact of ultrasonication on the effects of anaerobic digestion of dairy wastewater. The highest technological effects were recorded with sonification mode of 2 min work/28 min break. Increase in ultrasonic intensity significantly reduced the efficiency of organic compounds' removal from wastewater, as well as biogas yield.

Regardless of the experimental series, low nutrient removal efficiency was observed. The total phosphorus removal ranged from 2.04 ± 0.38 to 4.59 ± 0.68%, while the total nitrogen removal varied from 9.67 ± 3.36 to 20.36 ± 0.32%. The differences in biomass growth yield coefficients were not significant during the study and ranged from 0.12 ± 0.01 g/g COD removed to 0.19 ± 0.01 g/g COD removed.

The effects obtained in the study (referring to the efficiency of wastewater treatment, biogas production, as well as to the results of economic analysis) proved that an innovative multi-section horizontal flow reactor can be competitive to existing industrial technologies of food wastewater treatment.

Author Contributions: Conceptualization, M.D., M.Z., and J.K.; Data curation, M.D., M.Z., M.K., and J.K.; Formal analysis, M.D., M.Z., and J.K.; Funding acquisition, M.D. and M.Z.; Investigation, M.D., M.Z., and M.K.; Methodology, M.D., M.Z., M.K., and J.K.; Project administration, M.D., M.Z., and M.K.; Resources, M.D., M.Z., M.K., and J.K.; Software, M.D., M.Z., M.K., and J.K.; Supervision, M.D., M.Z., and M.K.; Validation, M.D., M.Z., and M.K.; Visualization, M.D., M.Z., and J.K.; Writing—original draft, M.D., M.Z., M.K., and J.K.; Writing—review & editing, M.D., M.K., and J.K. All authors have read and agreed to the published version of the manuscript.

Funding: The study was carried out in the framework of the statutory funds for research no. 29.610.023, financed by the Ministry of Science and Higher Education.

Conflicts of Interest: The authors declare no conflict of interest. The funders had no role in the design of the study; in the collection, analyses, or interpretation of data; in the writing of the manuscript, or in the decision to publish the results.

References

1. Ekanthalu, V.S.; Morscheck, G.; Narra, S.; Nelles, M. Hydrothermal Carbonization—A Sustainable Approach to Deal with the Challenges in Sewage Sludge Management. In *Urban Mining and Sustainable Waste Management*; Springer: Singapore, 2020; pp. 293–302.
2. Smol, M.; Adam, C.; Preisner, M. Circular economy model framework in the European water and wastewater sector. *J. Mater. Cycles Waste Manag.* **2020**, *22*, 682–697. [CrossRef]
3. Bot, F.; Plazzotta, S.; Anese, M. Chapter 16—Treatment of Food Industry Wastewater with Ultrasound: A Big Opportunity for the Technology. *Ultrasound Adv. Food Process. Preserv.* **2017**, 391–408. [CrossRef]
4. Bhatti, Z.A.; Maqbool, F.; Malik, A.H.; Mehmood, Q. UASB reactor startup for the treatment of municipal wastewater followed by advanced oxidation process. *Braz. J. Chem. Eng.* **2014**, *31*, 715–726. [CrossRef]
5. Schnürer, A.; Bohn, I.; Moestedt, J. Protocol for Start-Up and Operation of CSTR Biogas Processes. In *Hydrocarbon and Lipid Microbiology Protocols*; McGenity, T., Timmis, K., Nogales, B., Eds.; Springer Protocols Handbooks; Protocol for Start-Up and Operation of CSTR Biogas Processes; Springer: Berlin/Heidelberg, Germany, 2016; pp. 171–200.
6. Rajeshwari, K.V.; Balakrishnan, M.; Kansal, A.; Lata, K.; Kishore, V.V.N. State-of-the-art of anaerobic digestion technology for industrial wastewater treatment. *Renew. Sustain. Energy Rev.* **2000**, *4*, 135–156. [CrossRef]
7. Nelson, M.J.; Nakhla, G.; Zhu, J. Fluidized-Bed Bioreactor Applications for Biological Wastewater Treatment: A Review of Research and Developments. *Engineering* **2017**, *3*, 330–342. [CrossRef]
8. Goli, A.; Shamiri, A.; Khosroyar, S.; Talaiekhozani, A.; Sanaye, R.; Azizi, K. A Review on Different Aerobic and Anaerobic Treatment Methods in Dairy Industry Wastewater. *J. Environ. Treat. Tech.* **2019**, *6*, 113–141.
9. Leyva-Díaz, J.C.; Monteoliva-García, A.; Martín-Pascual, J.; Munio, M.M.; García-Mesa, J.J.; Poyatos, J.M. Moving bed biofilm reactor as an alternative wastewater treatment process for nutrient removal and recovery in the circular economy model. *Bioresour. Technol.* **2020**, *299*, 122631. [CrossRef]
10. Cravotto, G.; Binello, A.; Di Carlo, S.; Orio, L.; Wu, Z.L.; Ondruschka, B. Oxidative degradation of chlorophenol derivatives promoted by microwaves or power ultrasound: A mechanism investigation. *Environ. Sci. Pollut. Res.* **2010**, *17*, 674–687. [CrossRef]
11. Mishra, R.R.; Sharma, A.K. Microwave–material interaction phenomena: Heating mechanisms, challenges and opportunities in material processing-Review. *Compos. Part A* **2016**, *81*, 78–97. [CrossRef]
12. Rocky Flats Technology Summary. 2001. Available online: http://www.nttc.edu/env/Rock--Flats/RF--chap1.html (accessed on 25 March 2020).
13. Remya, N.; Lin, J.G. Current status of microwave application in wastewater treatment—A review. *Chem. Eng. J.* **2011**, *166*, 797–813. [CrossRef]
14. Kappe, C.O. Controlled microwave heating in modern organic synthesis. *Angew. Chem.* **2004**, *43–46*, 6250–6284. [CrossRef]
15. Gogate, P.R. Cavitation: An auxiliary technique in wastewater treatment schemes. *Adv. Environ. Res.* **2002**, *6*, 335–358. [CrossRef]
16. Matouq, M.A.D.; Al-Anber, Z.A. The application of high frequency ultrasound waves to remove ammonia from simulated industrial wastewater. *Ultrason. Sonochem.* **2007**, *14*, 393–397. [CrossRef] [PubMed]
17. Sangave, C.P.; Pandit, B.A. Ultrasound pre-treatment for enhanced biodegradability of the distillery wastewater. *Ultrason. Sonochem.* **2014**, *11*, 197–203. [CrossRef] [PubMed]

18. Blume, T.; Neis, U. Improved wastewater disinfection by ultrasonic pre-treatment. *Ultrason. Sonochem.* **2004**, *11*, 333–336. [CrossRef]
19. Kyllonen, H.; Pirkonen, P.; Nystrom, M.; Nuortila-Jokinen, J.; Gronroos, A. Experimental aspects of ultrasonically enhanced cross-flow membrane filtration of industrial wastewater. *Ultrason. Sonochem.* **2006**, *13*, 295–302. [CrossRef] [PubMed]
20. Seungmin, N.; Young-Uk, K.; Jeehyeong, K. Physiochemical properties of digested sewage sludge with ultrasonic treatment. *Ultrason. Sonochem.* **2007**, *14*, 281–285.
21. Sangave, C.P.; Pandit, B.A. Ultrasound and enzyme assisted biodegradation of distillery wastewater. *J. Environ. Manag.* **2006**, *80*, 36–46. [CrossRef]
22. Oz, N.A.; Uzun, A.C. Ultrasound pretreatment for enhanced biogas production from olive mill wastewater. *Ultrason. Sonochem.* **2015**, *22*, 565–572. [CrossRef]
23. Mischopoulou, M.; Naidis, P.; Kalamaras, S.; Kotsopoulos, T.A.; Samaras, P. Effect of ultrasonic and ozonation pretreatment on methane production potential of raw molasses wastewater. *Renew. Energy* **2016**, *96*, 1078–1085. [CrossRef]
24. Gude, V.G. *Microwave-Meditated Biofuel Production*, 1st ed.; CRC Press: Boca-Ratón, FL, USA, 2017.
25. Amini, A.; Aponte-Morales, V.; Wang, M.; Dilbeck, M.; Lahav, O.; Zhang, Q. Cost-effective treatment of swine wastes through recovery of energy and nutrients. *Waste Manag.* **2017**, *69*, 508–517. [CrossRef]
26. Ma, H.; Guo, Y.; Qin, Y.; Li, Y.Y. Nutrient recovery technologies integrated with energy recovery by waste biomass anaerobic digestion. *Bioresour. Technol.* **2018**, *269*, 520–531. [CrossRef] [PubMed]
27. Omil, F.; Garrido, J.M.; Arrojo, B.; Mendez, R. Anaerobic filter reactorperformance for the treatment of complex dairy wastewater AT industrial scale. *Water Res.* **2003**, *37*, 4099–4108. [CrossRef]
28. Mendez, R.; Blazquez, R.; Lorenzo, F.; Lema, J.M. Anaerobic treatment of cheese whey: Start-up and operation. *Water Sci. Technol.* **1989**, *21*, 1857–1860. [CrossRef]
29. Najafpour, G.D.; Hashemiyeh, B.A.; Asadi, M.; Ghasemi, M.B. Biological Treatment of Dairy Wastewater in an Upflow Anaerobic Sludge-Fixed Film Bioreactor American-Eurasian. *J. Agric. Environ. Sci.* **2008**, *4*, 251–257.
30. Danalewich, J.R.; Papagiannis, T.G.; Belyea, R.L.; Tumbleson, M.E.; Raskin, L. Characterization of dairy waste streams, current treatment practices and potential for biological nutrient removal. *Water Res.* **1998**, *32*, 3555–3568. [CrossRef]
31. Isik, M.; Sponza, D.T. Anaerobic/aerobic sequential treatment of a cotton textile mill wastewater. *J. Chem. Technol. Biotechnol.* **2004**, *79*, 1268–1274. [CrossRef]
32. Sklyar, V.; Epov, A.; Gladchenko, M.; Danilovich, D.; Kalyuzhnyi, S. Combined biologic (anaerobic–aerobic) and chemical treatment of starch industry wastewater. *Appl. Biochem. Biotechnol.* **2003**, *109*, 253–262. [CrossRef]
33. Ramasamy, E.V.; Gajalakshmi, S.; Sanjeevi, R.; Jithesh, M.N.; Abbasi, S.A. Feasibility studies on the treatment of dairy wastewaters with upflow anaerobic sludge blanket reactors. *Bioresour. Technol.* **2004**, *93*, 209–212. [CrossRef]
34. Raj, S.A.; Murthy, D.V.S. Comparison of the trickling filtermodels for the treatment of synthetic dairy wastewater. *Bioprocess Eng.* **1999**, *21*, 51–55. [CrossRef]
35. Göblös, S.; Portörő, P.; Bordás, D.; Kálmán, M.; Kiss, I. Comparison of the effectivities of two-phase and single-phase anaerobic sequencing batch reactors during dairy wastewater treatment. *Renew. Energy* **2008**, *33*, 960–965. [CrossRef]
36. Kotsyurbenko, O.R.; Friedrich, M.W.; Simankova, M.V.; Nozhevnikova, A.N.; Golyshin, P.N.; Timmis, K.N.; Conrad, R. Shift from acetoclastic to H2-dependent methanogenesis in a West Siberian peat bog at low pH values and isolation of an acidophilic Methanobacterium strain. *Appl. Environ. Microbiol.* **2007**, *73*, 2344–2348. [CrossRef] [PubMed]
37. Gabriel, C.; Gabriel, S.; Grant, E.H.; Halstead, B.S.J.; Mingos, D.M.P. Dielectric parameters relevant to microwave dielectric heating. *Chem. Soc. Rev.* **1998**, *27*, 213–223. [CrossRef]
38. Banik, S.; Sharma, A.; Dan, D. Effect of microwave on cell biology of Methanosarcina Barkeri. *J. Mycopathol. Res.* **2003**, *41*, 217–219.
39. Parker, M.C.; Besson, T.; Lamare, S.; Legoy, M.D. Microwave radiation can increase the rate of enzyme catalysed reaction in organic media. *Tetrahedron Lett.* **1996**, *37*, 8383–8386. [CrossRef]

40. Sanjay, G.; Sugunan, S. Enhanced pH and thermal stabilities of invertase immobilized on montmorillonite K-10. *Food Chem.* **2006**, *94*, 573–579. [CrossRef]
41. Fang, Y.; Sun, S.-Y.; Xia, Y.-M. The weakened 1,3-specificity in the consecutive microwave assisted enzymatic synthesis of glycerides. *J. Mol. Catal. B Enzym.* **2008**, *55*, 6–11. [CrossRef]
42. Ahn, J.H.; Shinb, S.G.; Hwangb, S. Effect of microwave irradiation on the disintegration and acidogenesis of municipal secondary sludge. *Chem. Eng. J.* **2009**, *153*, 145–150. [CrossRef]
43. Eskicioglu, C.; Kennedy, K.J.; Droste, R.L. Characterization of soluble organic of waste from sewage sludge before and after thermal pretreatment. *Water Res.* **2006**, *40*, 3725–3736. [CrossRef]
44. Oz, N.A.; Yarimtepe, C.C. Ultrasound assisted biogas production from landfill leachate. *Waste Manag.* **2014**, *34*, 1165–1170. [CrossRef]
45. Yarimtepe, C.C.; Oz, N.A. Investigation of the pretreatment effect of ultrasound on anaerobic sequencing batch reactor treating landfill leachate. WIT Transactions on Ecology and the Environment. *Water Resour. Manag. IX* **2017**, *220*, 93–98.
46. Nivedha Ramanathan, R.M.; Balasubramanian, N.; Chithra, K. Biogas from confectionery wastewater with the application of ultrasound pre-treatment. *Energy Sources Part A Recovery Util. Environ. Effects* **2019**. [CrossRef]
47. Kavitha, S.; Rajesh Banu, J.; Kumar, G.; Kaliappan, S.; Yeom, I.T. Profitable ultrasonic assisted microwave disintegration of sludge biomass: Modelling of biomethanation and energy parameter analysis. *Bioresour. Technol.* **2018**, *254*, 203–213. [CrossRef] [PubMed]
48. Yin, X.; Lu, X.; Han, P.; Wang, Y. Ultrasonic treatment on activated sewage sludge from petro-plant for reduction. *Ultrasonics* **2006**, *44*, e397–e399. [CrossRef] [PubMed]
49. Witkowska, E.; Buczkowska, A.; Zamojska, A.; Szewczyk, K.W.; Ciosek, P. Monitoring of periodic anaerobic digestion with flow-through array of miniaturized ion-selective electrodes. *Bioelectrochemistry* **2010**, *80*, 87–93. [CrossRef]
50. Begum, S.; Anupoju, G.R.; Sridhar, S.; Bhargava, S.K.; Jegatheesan, V.; Eshtiaghi, N. Evaluation of single and two stage anaerobic digestion of landfill leachate: Effect of pH and initial organic loading rate on volatile fatty acid (VFA) and biogas production. *Bioresour. Technol.* **2018**, *251*, 364–373. [CrossRef] [PubMed]
51. Seghezzo, L.; Zeeman, G.; Van Lier, J.B.; Hamelers, H.V.M.; Lettinga, G. A review: The Anaerobic treatment of sewage in UASB and EGSB reactors. *Bioresour. Technol.* **1998**, *65*, 175–190. [CrossRef]
52. Ozay, Y.; Ünşar, E.K.; Işık, Z.; Yılmaz, F.; Dizge, N.; Perendeci, N.A.; Mazmanci, M.A.; Yalvac, M. Optimization of electrocoagulation process and combination of anaerobic digestion for the treatment of pistachio processing wastewater. *J. Clean. Prod.* **2018**, *196*, 42–50. [CrossRef]
53. Gerek, E.E.; Yılmaz, S.; Koparal, A.S.; Gerek, Ö.N. Combined energy and removal efficiency of electrochemical wastewater treatment for leather industry. *J. Water Process Eng.* **2019**, *30*, 100382. [CrossRef]
54. Hiwarkar, A.D.; Singh, S.; Srivastava, V.C.; Mall, I.D. Mineralization of pyrrole, a recalcitrant heterocyclic compound, by electrochemical method: Multi-response optimization and degradation mechanism. *J. Environ. Manag.* **2017**, *198*, 144–152. [CrossRef]
55. Ribera-Pi, J.; Badia-Fabregat, M.; Calderer, M.; Polášková, M.; Svojitka, J.; Rovira, M.; Jubany, I.; Martínez-Lladó, X. Anaerobic Membrane Bioreactor (AnMBR) for the Treatment of Cheese Whey for the Potential Recovery of Water and Energy. *Waste Biomass Valorization* **2020**, *11*, 1821–1835. [CrossRef]
56. Masłoń, A. Energy Consumption of Selected Wastewater Treatment Plants Located in South-Eastern Poland. *Eng. Prot. Environ.* **2017**, *20*, 331–342. [CrossRef]
57. Longo, S.; d'Antoni, B.M.; Bongards, M.; Chaparro, A.; Cronrath, A.; Fatone, F.; Lema, J.M.; Mauricio-Iglesias, M.; Soares, A.; Hospido, A. Monitoring and diagnosis of energy consumption in wastewater treatment plants. A state of the art. And proposals for improvement. *Appl. Energy* **2016**, *176*, 1251–1268. [CrossRef]

 *energies*

Article

Can Multiple Uses of Biomass Limit the Feedstock Availability for Future Biogas Production? An Overview of Biogas Feedstocks and Their Alternative Uses

Dieu Linh Hoang [1,*], Chris Davis [2], Henri C. Moll [1] and Sanderine Nonhebel [1]

[1] Integrated Research on Energy, Environment and Society (IREES), Nijenborgh 6, 9747 AG Groningen, The Netherlands; h.c.moll@rug.nl (H.C.M.); s.nonhebel@rug.nl (S.N.)
[2] Invenia Labs, 95 Regent Street, Cambridge CB2 1AW, UK; chris.davis@invenialabs.co.uk
* Correspondence: d.l.hoang@rug.nl; Tel.: +31-50-363-9826

Received: 27 April 2020; Accepted: 22 May 2020; Published: 30 May 2020

Abstract: Biogas is expected to contribute 10% of the total renewable energy use in Europe in 2030. This expectation largely depends on the use of several biomass byproducts and wastes as feedstocks. However, the current development of a biobased economy requires biomass sources for multiple purposes. If alternative applications also use biogas feedstocks, it becomes doubtful whether they will be available for biogas production. To explore this issue, this paper aims to provide an overview of potential alternative uses of different biogas feedstocks being researched in literature. We conducted a literature review using the machine learning technique "co-occurrence analysis of terms". This technique reads thousands of abstracts from literature and records when pairs of biogas feedstock-application are co-mentioned. These pairs are assumed to represent the use of a feedstock for an application. We reviewed 109 biogas feedstocks and 217 biomass applications, revealing 1053 connections between them in nearly 55,000 scientific articles. Our results provide two insights. First, a large share of the biomass streams presently considered in the biogas estimates have many alternative uses, which likely limit their contribution to future biogas production. Second, there are streams not being considered in present estimates for biogas production although they have the proper characteristics.

Keywords: biogas; biomass waste; competing uses; biomass applications; bio-based economy; biomass value pyramid; co-occurrence analysis

---

1. Introduction

Interest in biogas has been growing in recent years, particularly due to its potential to contribute to a renewable energy transition. Through anaerobic digestion, biomass is converted into biogas containing methane. This biogas could be used as a substitute for natural gas in providing electricity, heat and methane for chemical uses [1].

In policy documents on European future energy supply, biogas plays an important role. The Directive 2018/2001 promotes biogas as one of the strategic means for the European Union (EU) (*EU's data in this paper includes United Kingdom*) to reach the target of 32% renewable energy use in 2030 [2]. In 2030, the expected availability of biogas is 1.7 EJ, which is equivalent to 3.7% of the energy consumption of the EU. To create such a biogas contribution, one-third is derived from energy crops, while the rest is mainly produced from manure and different types of organic waste [3,4]. In other research, straws, sewage sludge, organic house waste, food waste, waste from landscape management and byproducts from the food and beverage industries are also considered as potential major sources for European biogas production in the coming period [5–7].

However, because of several grounds, these expectations may lead to over-estimations of future biogas production. First, the use of energy crops is under heavy discussion since it competes with food and feed [8,9] and the production of these crops have large environmental impacts [10–12]. For these reasons, energy crops are less likely to be widely used for biogas production. Biomass from byproducts and waste comes from existing streams and is generally considered more sustainable for biogas production [11,13,14]. Second, the estimates for future waste stream availability for biogas production are based on the present situation. In the coming decades it is likely that energy and material use will change drastically in society. One change is the movement towards a biobased economy. This implies that biobased resources will be used differently in society. For example, there have been researches on using sewage sludge as a soil amendment [15]; straws in the production of several biofuels [16] and bioplastics [17]; and food waste in biopesticide production [18,19]. In such a biobased society, tough competition could be expected between different applications of byproducts and waste streams. The potential to produce biogas may be severely affected by this competition.

By analyzing the future competition around byproducts and waste streams, a better base for expectations of the future biogas production could be developed. The "Biomass Value Pyramid" (BVP) [20,21] (see Figure 1) offers a useful way to structure the analysis of the competition. The BVP is used already often in the discourse on multiple uses of a source. The pyramid indicates which applications can add the highest value to the biomass. The higher in the pyramid, the higher value of the applications as well as the more knowledge and skills are required for using the biomass [21]. With more economic value rewarded, the higher applications are likely to outcompete applications lower on the BVP such as energy which is at the bottom of the pyramid. Since biogas is primarily used for energy, it is also at the bottom which means it will have difficulty obtaining biomass feedstocks that have alternative uses.

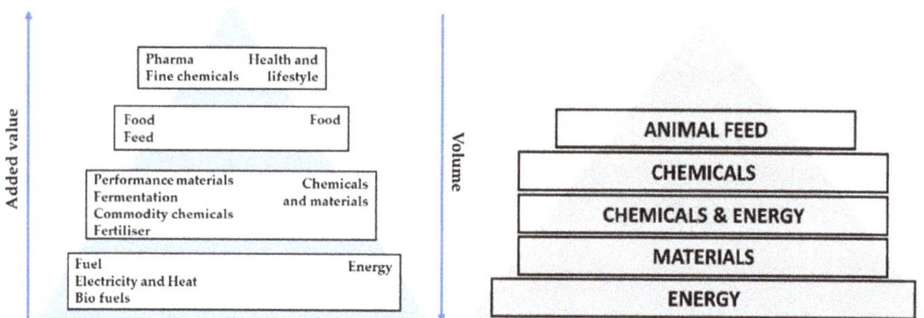

Figure 1. Biomass value pyramid [22] and biomass application categorization used in this research. *The pyramid indicates which applications can add the highest value to the biomass. The higher in the pyramid, the higher the value of the applications as well as the more knowledge and skills required for using the biomass.*

As mentioned earlier, a shift towards a more biobased society will lead to other uses of biomass than the present which likely affects the availability for biogas production. For realistic estimates for the future availability of biomass for biogas production these new future uses of biomass should be taken into account. Novel uses take time to be developed, so we assume that alternative uses of feedstocks in the future can be found in the scientific literature. To better understand the feedstocks competition for biogas production, an overview should be available identifying all biogas feedstocks and connecting these to all potential biomass applications taking into account the BVP. This would enable the identification of the feedstocks which are more likely to be available for biogas.

Currently, such an overview is lacking. This is a complicated task as it involves reviewing thousands of papers divided into many different research niches. Effectively processing this amount of information would be unfeasible without automated assistance.

In this paper, through a machine learning literature review method, we aim to provide an overview of potential alternative uses of different biogas feedstocks being researched in nearly 55,000 scientific papers. In this way, we seek to develop better knowledge about the role of potential biomass competition in the biobased economy, especially for the contribution of biogas production in the future energy transition.

2. Materials and Methods

2.1. Machine Learning Approach: The Co-Occurrence Analysis of Terms

Machine learning is referred to as "the automated detection of meaningful patterns in data" [23]. This approach combines multiple techniques to extract information from large data sets [23]. In this research, we use the technique "co-occurrence analysis of terms" described by Davis et al. [22]. This technique can be used to read abstracts of thousands of documents, scan these for terms specified by the researcher, and then record co-occurrences of these terms [22]. This work only examines co-occurring terms if they appear in separate pre-defined lists. By making one list of terms describing feedstocks and another with terms describing biomass applications, this technique will identify pairs of feedstocks and applications mentioned together in literature.

In [22], Davis and his colleagues also used this technique on a list of biomass terms and a list of application terms to identify value pathways for organic wastes. The results provide almost 2500 connections between 450 organic wastes and 200 applications which is much broader than covered by previous review studies. Although the authors also pointed out that the method did not "extract the nature of relationship between the terms" [22], it does show the potential of this technique to uncover combinations of feedstocks and biomass that research may not immediately think of.

To provide an overview of the alternative uses for different biogas feedstocks found in literature, we adapted this technique for our research. We utilized the co-occurrence analysis algorithm provided by Davis, although we improved the list of feedstocks to make it more focused on biogas feedstocks. In addition, we also improved the list of applications with a better structured categorization as Davis [22] suggested.

Following the guidelines of Davis [22], our research starts with three steps: *(1) literature collection; (2) identifying biogas feedstocks and biomass applications; (3) co-occurrence calculation*. Steps 1 and 2 are manual collections which will be described in details in Section 2.2. Step 3 is the use of the co-occurrence calculation algorithm of Davis [22] in R [24]. Source codes of the first three steps are online and accessible via the link mentioned in Supplementary Materials Appendix D. The main results of these steps are as follows:

i. A co-occurrence matrix: We use one large grid matrix to demonstrate which biogas feedstock is mentioned along with certain applications. Here we define a co-occurrence as one pair of a specific feedstock and application which are co-mentioned at least once in the literature (e.g., maize stover and bioethanol, pig manure and compost). If there is at least one co-occurrence in the included literature, the connection between the biomass and the application is shown in the matrix. The order of the rows and columns (feedstocks and applications) in the matrix is based on the similarity of the terms in the literature collection. It presents similar types of feedstocks suitable for an application near each other and similar types of applications requiring a feedstock are also near each other. This ordering is done in two steps: first there is an automated ordering based on hierarchical clustering of the row and column values in the original matrix output by the algorithm [22]; second, we manually check and adjust the ordering.

ii. A list of literature: This indicates the exact literature where each co-occurrence happened.

Furthermore, we are aware that the result of the "co-occurrence analysis of terms" algorithm by Davis [22] is simply the co-mentioning of each word pair and does not explicitly indicate the relation between these terms. Therefore, we performed the fourth step *(4) co-occurrence validation* which strengthens the interpretation from the co-occurrences to the potential alternative uses of the biogas feedstocks. This step is described in more detail in Section 2.3. See Figure 2 for the overview of the research methodology.

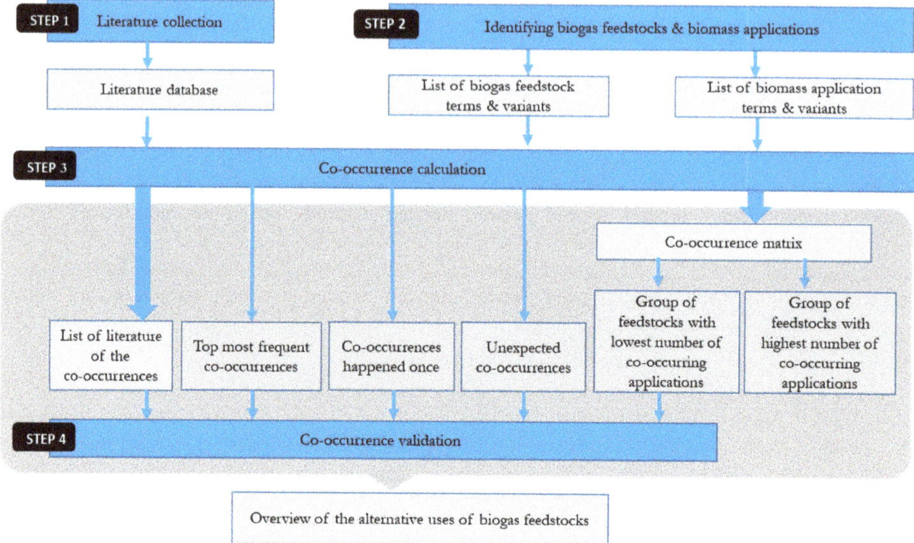

Figure 2. Overview of the methodology.

2.2. Collecting Literature on Biomass Applications

The aim of this step is to collect a large number of articles which discuss the applications of biomass. We ran the search query mentioned below on ScienceDirect. ScienceDirect is a large scientific publication database whose topics vary from Physical Sciences and Engineering, Life Sciences, Health Sciences to Social Sciences and Humanities [25]. This large and broad-scope collection, and the fact that the database allows for automatically downloading thousands of abstracts [26] makes ScienceDirect suitable for collecting literature for this research.

- Search query: *queryString = "title-abs-key* (technology *OR* process *OR* conversion *OR* treatment *OR* use *OR* production *OR* application) *AND title-abs-key* (product *OR* waste *OR* by-product *OR* byproduct *OR* feedstock *OR* additive *OR* catalyst) *AND title-abs-key* (organic *OR* bio *OR* biomass)".

2.3. Identifying Biogas Feedstock and Biomass Application Terms

2.3.1. Identifying Biogas Feedstocks

The list of biogas feedstocks used in this research is mainly based on the guidance of the Food and Agriculture Organization (FAO) Biogas Industrial User Manual [27,28] and the European Feedstock Atlas [29,30]. The lists of biogas feedstocks suggested by these two reports are quite close to each other. They also cover a large variety of byproducts and waste biomass which are potentially usable for European biogas production and have been discussed in literature. In these reports, the feedstocks are grouped into different categories reflecting the type of process producing them. For example, wheat straw belongs to the category of Crop leftovers and coconut extraction meal belongs to the category of Vegetable oil production.

Each biogas feedstock recommended by FAO and European Feedstock Atlas is a unique stream, which differs from another by its own biogas yield, physical and chemical characteristics such as dry matter and volatile solids content. This is understandable because these are technical documents.

However, in our research, it is important to define the terms which are likely to appear in the literature. We identified four confusing situations and decided to group certain feedstocks to address this:

- A generic stream and a specific stream: There are some streams which are general but seem to overlap with other streams such as dairy industry waste which may conceptually include cheese waste. However, in the context of their applications they are different streams. So, if the two main lists mentioned both, we will keep both the generic and the specific terms.

- Streams with names which may be interchangeable in literature but have very different physical properties according to the two main lists, such as beer barm and brewer's grains. We decided to keep them as two separate terms in our list.

- Streams with names which may interchangeable in literature and have similar physical properties according to the two main lists (e.g: maize straw and maize stover) or just different in the water content (e.g., cattle slurry and cattle manure). We decided to combine them into one term.

- Streams which are rarely found in the scientific literature on biomass applications such as "barley feeding meal". We excluded these from our lists.

To verify whether the terms are similar, or to understand more about the feedstocks, we have to compare them in Google, different literature and several websites about feedstocks such as Feedipedia [31] and Feedbase [32]. Additionally, we added biomass sources mentioned in [33] and [34] which were not included in FAO Biogas Industrial User Manual and the European Feedstock Atlas.

Due to the criticisms mentioned in the introduction, energy crops are excluded in this research.

2.3.2. Identifying Biomass Applications

The Biomass Value Pyramid (BVP) categorizes the final uses of biomass into groups based on their relative economic value. The detailed categorizations vary from literature but these six groups are usually mentioned, from high to low value: Pharmaceuticals, Human Food, Animal Feed, Chemicals, Materials and Energy [20,21,35]. This gives a useful direction on which literature we should look at to inventory the specific applications. With the inspiration of the BVP, we categorized the specific applications into five groups: Animal Feed, Chemicals, Chemical and Energy, Materials and Energy. "Food" is omitted as we only considered byproducts and waste streams. "Pharmaceuticals and high-end chemicals" is merged with "Chemicals", because biomass requires chemical processing before being used in these higher value applications. This is part of a broader issue as for many chemicals as it is almost impossible to determine the end application of a chemical substance. Additionally, we distinguished the group Chemical and Energy because some chemicals such as biomethane and biohydrogen can be used for both energy or further chemical industrial processes (see Figure 1). This means that the group "Energy" only includes applications which are not also used as chemicals.

To define the terms for each application, we have different strategies for each category:

- Animal feed: The feedstocks that can be fed to animals raw or after processing. Although the processing consists of multiple steps with intermediate components, we only cared whether those components led to animal feed or not. As a result, we grouped terms found in literature such as forage, fodder, and animal feed supplements into a single term: animal feed.

- Chemicals: We used the original terms used in the approach from Davis et al. [22] supplemented by terms from other literature.

- Materials: We used the existing material terms mentioned by Davis et al. and added biobased materials mentioned in other bioeconomy literatures such as different types of bioplastics.

- Energy: Unlike chemical and materials which have multiple conversion processes, bioenergy has quite few and straightforward conversion processes. In literature, energy applications can be described by either end use products or the conversion processes, for example, torrefied biomass and torrefaction. We did an inventory of different bioenergy processes to make sure that we did not miss any bioenergy applications. Then the terms for energy applications were defined by their end products.

- Chemicals and Energy: Like energy, the terms included in this group have few conversion processes and can be described by either end use products or conversion processes. We defined the terms for chemical and energy applications using the end products.

Multiple references used for identifying the list of biomass applications are mentioned in Supplementary Materials Appendix B.

2.3.3. Variants of the Terms

As mentioned above, a specific feedstock or application can be described by several synonyms. In processing the literature to identify the co-occurrences, we consider all of these synonyms and also other term variants such as plural and singular forms, and differing adjectives. in the presentation of the result, we use a single term for each biogas feedstock and each biomass application. As a part of the technique from Davis et al. [22], there is an algorithm to help generate the variants of our considered terms in the literature collection and to group the result into one single term. It should be noted that while for other categories the term variants are only defined by the end products, the variants for Energy and Energy and Chemical applications are decided by both the end products and conversion processes.

2.4. Co-Occurrence Validation

Our overview of potential alternative uses of biogas feedstocks depends on the assumption that a co-occurrence represents the connection where a feedstock is used as input for an application. However, a feedstock and an application can be mentioned together for other reasons, especially when the application is referred to as an end product. For example, the co-mentioned end product is a substance required to process the feedstock. Therefore, in this research, we manually checked a sample of the co-occurrences to see to which extent the co-occurrences of our dataset reflect the assumed connection. We performed this manual check for four types of co-occurrence:

i. The most frequent 20 co-occurrences which happened in the literature collection;
ii. Random 20 co-occurrences which happened once in the literature collection: from the list of co-occurrences that happened once which was ordered alphabetically based on the feedstock terms, we first randomly chose the first co-occurrence to check. Then we checked the co-occurrence 20 below the previous one. This procedure was followed to collect 20 co-occurrences. However, if one of the collected co-occurrences was too similar to an earlier collected co-occurrence, we replaced it with the co-occurrence below it from the list;
iii. Random 20 co-occurrences which are unexpected to the knowledge of the authors. For example, we checked co-occurrences where feedstocks are considered as waste and the applications belong to higher level of the BVP. The list of co-occurrences was ordered alphabetically based on the feedstock terms;
iv. All co-occurrences of the feedstocks which have the lowest number of co-occurring applications.

The first two validations were done to check whether a co-occurrence found frequently in the literature more reliably represents the use of a feedstock for an application than those found in only one. The third validation was to see whether the unexpected co-occurrences includes the one that represent our assumption. The fourth validation was performed because the interpretation of feedstocks with very few co-occurring applications is likely more sensitive to mistaken co-occurrences than feedstocks with many co-occurring applications.

To validate each co-occurrence, we used the list of literature (Supplementary Materials Appendix E), a product of the co-occurrence calculation algorithm, to locate the exact literature(s) where the concerned co-occurrences happened. Then we read the abstract(s) to understand the connection between the feedstock and the application. The validation stops when we find one abstract from literature confirming the assumed connection. If none of the located literature confirms the assumed connection, we consider it a false positive.

3. Results

3.1. Results of Literature Collection and Identifying Biogas Feedstocks and Biomass Applications

We collected the abstracts of 54,322 distinct articles about the applications of biomass from ScienceDirect. The articles are scientific research from the years 1970 to 2018. We also identified 109 biogas feedstocks which are byproducts and wastes streams, and 217 biomass applications. These are the inputs for generating the co-occurrence matrix described in 3.3. Table 1 shows the energy application terms and their conversion process which are used to generate more variants of terms to capture this application in the literature. Detailed lists of application and feedstocks, their term variants and references are presented in Supplementary Materials Appendices A and B.

Table 1. Energy application terms and their conversion process summarized by this research.

	Type of Bioenergy		Conversion Processes	Common Feedstocks
	Common Name [1]	Other Names		
Liquid fuels	Biodiesel	*Biodiesel*	transesterification	oil crops feedstocks, animal fats
		Syndiesel	gasification + Fischer Tropsch	cellulose feedstocks
		Renewable diesel	hydrogenation	oil crops feedstocks, animal fats
		HTU diesel	hydro thermal upgrading (HTU)	wet feedstocks (e.g., beet pulp)
	Bio gasoline		gasification + Fischer Tropsch	wood waste, algae
	Bio jet fuel	*Bio jet fuel*	alcohol-to-jet	various feedstocks
			oil-to-jet	various feedstocks
			sugar-to-jet	various feedstocks
		Synthetic jet fuel	gas-to-jet	cellulose feedstocks
	Bioethanol [2]		sugar fermentation	sugar and starch feedstocks
			hydrolysis + fermentation	cellulose feedstocks, algae
			gasification + fermentation	various feedstocks, waste
	Pyrolysis oil [2]		pyrolysis	wood, algae
			hydrothermal Liquefaction	algae
	Direct vegetable oil [2]	*New oil* [3]	oil pressing/oil extraction	oil crops feedstocks
		Recycled oil	-	waste vegetable oil
	Dimethyl ether [2]		gasification + dehydration (of methanol)	black liquor, cellulose feedstocks
			gasification + synthesis (with CH_4 from anaerobic digestion)	various feedstocks, waste
	Methanol [2]		gasification + synthesis (from CO, CO_2, H_2)	various feedstocks
			biosynthesis (from CH_4)	
	Butanol [2]		ABE fermentation	cellulose feedstocks, sugar and starch feedstocks, algae
	Others [2]			
	Bio-ethers [2]		chemical synthesis	-
	Dimethylfuran [2]		chemical synthesis	-

Table 1. *Cont.*

Type of Bioenergy		Conversion Processes	Common Feedstocks
Common Name [1]	Other Names		
Gaseous fuels			
Biomethane [4]	*biogas*	anaerobic digestion	various feedstocks
	biomethane	gasification + synthesis	cellulose feedstocks
Biohydrogen [2]		dark fermentation	carbohydrates
		photobiological hydrogen production	algae, waste water
Syngas [2]		gasification	cellulose feedstocks
Solid fuels			
Use biomass as it is		combustion, firing	cellulose feedstocks, manure
		incineration	various feedstocks
Briquette		desification, pelletization, briquetting	cellulose feedstocks, manure
Torrefied biomass		torrefaction	cellulose feedstocks

[1] defined as application terms in this research; [2] also has Chemical applications; [3] not from byproducts and waste streams, therefore, not discussed in this research; [4] not included in the application list since gasification is included in syngas as an end product and all the feedstocks in this research are verified to be possible to use for biogas production.

3.2. Appeared Terms in the Co-Occurrences, Co-Occurrence Validation and Unexpected Connections

Two-third of the feedstocks and applications which were identified in step two showed up in the co-occurrences: 71/109 biogas feedstocks and 150/217 applications. This means that the 150 applications and the 71 feedstocks have in some way been mentioned together in the literature collection. The remaining feedstocks and applications might be mentioned individually in the literature collection but not with another term on the lists to form a pair of feedstock-application co-occurrences. Some feedstocks which did not appear in the co-occurrences have similar physical properties with the ones which appeared. For example, wheat straw appeared in the co-occurrences but oat straw did not. All categories of feedstocks and applications have their representatives in the co-occurrences.

The pairing between the feedstocks and applications which appeared resulted in 1053 co-occurrences. Half of the co-occurrences only happened once in literature. Of these, 102 co-occurrences were manually checked according to the four criteria mentioned. Among the checked subset, in total, 65% of the co-occurrences represented the assumed connection that the feedstock is an input for the co-occurring applications. This percentage varied between the four validated groups: 95% for the most frequent co-occurrences; 40% for the co-occurrences happened once; 58% for the unexpected co-occurrences; and 66% for the co-occurrences of the feedstock group with lowest number of co-occurring applications. On one hand, the first two show that co-occurrences which happened once are more likely to be mistaken than those that happened more frequent. On the other hand, 40% is also a substantial number that we would miss if we disregard those were found once. The last two show that despite large amounts of false positives, these two groups include sizable numbers of co-occurrences which indeed represent the assumed connections.

From the co-occurrences that did not match our assumed connection, some appeared frequent in the literature collection and some belonged to the unexpected group. For the high frequent co-occurrences, the false positive one that we observed often appear in the literature with a quite consistent connection. Glycerol and acetic acid are often mentioned together because they are often mixed together in common chemical processes [36,37]. In the group of unexpected co-occurrences which mismatch with our assumed connection also, most are mixed wastes so what we see is that the application terms actually indicate the other components of the waste stream. For example, organic house waste co-occurred with different types of bioplastics [38–40].

However, the co-occurrences also reveal unexpected connections which truly represent our assumed connection. For example, "rice straw", a cellulose-rich material, can be used in fatty acids extraction; one article describes an engineering experiment that successfully extracted fatty acids from a type of micro bacteria grown rice straw substrate for biofuel production in a similar manner to algae oil/microbial oil [41]. Another example is that different types of manure can be turned into biodiesel, pyrolysis oil, acetone, and methyl esters [42–45]

For more details on co-occurrence validation, see Supplementary Materials Appendix C.

3.3. Co-Occurrence Matrix

In the co-occurrence matrix (Figure 3 and Supplementary Materials Appendix D), each row represents one biogas feedstock and each column represents one biomass application. The black cell at the meeting point of one row and one column indicates that the feedstock and the application co-occurred at least once in the literature collection. White cells mean that the feedstock and the application does not co-occur in the literature collection. We can see several clusters in the matrix.

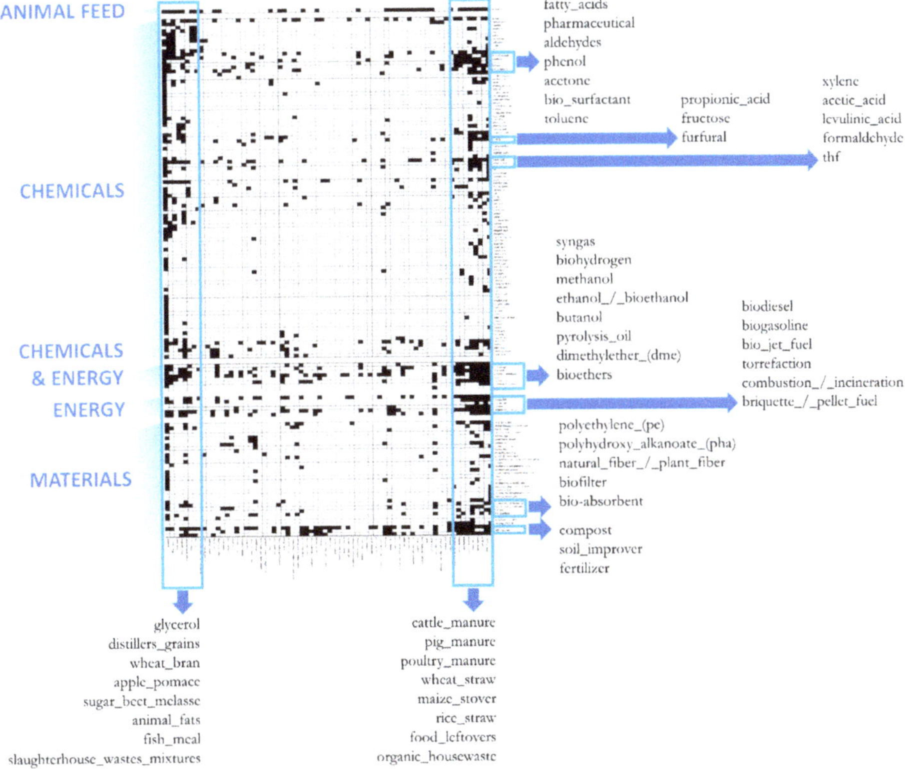

Figure 3. The co-occurrence matrix: overview of the co-occurrences between biogas feedstocks and biomass applications in the literature collection in this research. *Each row represents one biogas feedstock and each column represents one biomass application. The black cell at the meeting point of one row and one column indicates that the feedstock and the application co-occurred at least once in the literature collection. A white cell means that the feedstock and the application does not co-occur in the literature collection.*

The feedstock groups at the two ends of the horizontal axis co-occurred with many applications across all application categories. The first group consists of homogenous streams with high contents of a specific substance such as a lipid (glycerol, animal fats, fish meal), protein (distilled grains), fiber (maize stover, wheat straw), sugar (sugar beet molasses), or nitrogen (different types of manure). The second group consists of mixed streams like organic waste and food leftover. This could be expected since feedstocks with a high concentration of a desired substance makes it easier to extract, and mixed streams might contain multiple desirable substances to extract.

With regards to the applications, animal feed and the group of fertilizer, soil improver and compost co-occurred with one-fourth of biogas feedstock streams. Different types of bio-energy except bio-ethers and bio jet fuel also co-occurred with many biogas feedstocks. Besides, certain types of materials and chemicals have other noticeable applications which appeared in a lot of the co-occurrences.

3.4. Top Biogas Feedstocks Having Highest and Lowest Number of Co-Occurring Applications

From results of the co-occurrence calculation, the number of applications that a feedstock co-occurred with ranges from 0–110. Almost half of the reviewed feedstocks have at least four co-occurring applications. We considered feedstocks co-occurring with at least 20 applications as having a high number of co-occurring applications; feedstocks co-occurring with three or fewer applications are considered to have a low number of co-occurring applications.

The calculation shows that 17/109 suggested feedstocks have co-occurrences with at least 20 applications. (see Table 2). These feedstocks do not only have a high number of co-occurring applications in total but also in each individual category. This means that these biogas feedstocks are likely to have multiple-alternative uses with a higher BVP level.

Table 2. Biogas feedstocks with the highest number of co-occurring applications in the literature collection in this research. Gray cells indicate that the corresponding feedstock belongs to the top co-occurrences of each category.

Feedstocks	Number of Applications on Category of					Total
	Animal Feed	Chemicals	Chemicals and Energy	Energy	Materials	
Glycerol	1	77	5	6	21	110
organic housewaste	1	35	5	7	18	66
wheat straw	1	33	4	7	15	60
sugar beet melasse	1	31	3	5	10	50
maize stover	1	27	4	8	9	49
food leftovers	1	22	5	8	9	45
rice straw	1	22	5	7	9	44
distiller's grains	1	25	3	5	3	37
cattle manure	0	13	4	4	8	29
wheat bran	1	25	2	0	1	29
malt coffee marc	1	8	4	4	10	27
pig manure	1	11	3	5	6	26
used paper	0	10	2	3	9	24
citrus bagasse	1	10	1	3	6	21
grape marc	1	8	2	3	6	20
poultry manure	1	7	2	5	5	20
animal fats	0	11	2	4	3	20

With regards to the group of feedstocks with the least number of co-occurring applications, 41 feedstocks were not co-mentioned with any applications in the literature and 16 feedstocks had 1–3 applications (see Table 3). Through the validation of this type of co-occurrences, we observed many applications of these feedstocks still belong to the chemicals and chemicals-energy high value group (e.g., biohydrogen, bioethanol, biodiesel).

Table 3. Biogas feedstocks with the lowest number of co-occurring applications in the literature collection in this research. The list is created after the validation step.

Feedstocks Co-Occurred with 1–3 Applications	Feedstocks with no Application Co-Occurrences		
natural grass	roadside grass	rabbit manure	apricot fiber
brewer's grains	distiller's residues	beer barm	succory waste
brewer's yeast	flax extraction meal	hemp press cake	asparagus waste
coconut extraction meal	maize germ extraction meal	safflower extraction meal	cauliflower residues
sunflower extraction meal	palm-kernel extraction meal	sesame extraction meal	eggplant residues
cotton seed extraction meal	peanut extraction meal	barley bran	turnip residues
sunflower peelings	oat straw	oat bran	spinach waste
rye straw	pea straw	rye bran	fruit industry wastewater
ryegrass straw	meadow hay	coffee industry wastewater	potatoes wastewater
cut flower wastes	slaughterhouse wastewater	dairy industry wastewaters	pulp and paper industry wastewater
brewery industry wastewaters	oilseed processing wastewater	slaughterhouse flotate fat	bakery residues
meat waste	rumen content	animal offals	old bread
animal bloodmeal	fish liver meal	animal hairmeal	
dairy industry wastes (general)	milk whey	cheese waste	
feeding beet leaves	sugar beet scraps		
olive pulp			

4. Discussion

The goal of this research was to provide an overview of alternative uses of biomass streams suitable for biogas production. This overview should provide insights in whether biomass streams are likely to be available for future biogas production or if competition with other uses will emerge. To do so we used a machine learning technique which made it possible to review nearly 55,000 papers. The quality of this review deviates from a normal review where the scientists reads the paper.

4.1. Limitations and Recommendations

There are two main limitations of the methodology in this research. On one hand, the overview includes co-occurrences which do not represent the assumed connection. Our validation identified false positives for a subset of co-occurrences. Having false positives will result in the alternative uses of feedstocks identified with the co-occurrence analysis to be higher than the actual number of uses. Improving the precision of the term variants will likely decrease this percentage of false positive, but it is unlikely to come close to 0 because co-occurrences with other strong connections will still show up. To have an overview without any false positive would likely require additional support from techniques which could predetermined the unwanted connections and filter them out from the result of co-occurrence calculation.

On the other hand, there are alternative uses of feedstocks which have been researched in literature that were not identified in our result. For example, oat straw had no co-occurrences despite the fact that it can be composted [46]. Such false negatives mean that feedstocks would have more actual uses than identified by our method. The false negative can be caused by two reasons. One is that the literature discussing these connections are not included in the database. Second is that the term variants are not precise enough to capture this connection. So, the overview presented is not covering all relations. This means that if one is interested in the use of one specific biomass stream extra literature search has to be done with other search queries.

The third limitation is about the scope of this study. In this research we only consider the current biogas feedstocks and study their potential use in the future. However, another aspect of the bio-economy is that due to the new way of using materials, there will be new biomass streams that can also be used for biogas production. There is research on the potential use for biogas of those waste streams [47,48]. However, there are no overviews of these new potential streams. To have more insight in the availability of biomass for biogas in the future, we recommend to have more research on these future streams and their alternative uses.

4.2. Robustness of the Research

The abovementioned limitations have an impact on the exact number of alternative uses of the biogas feedstocks in our overview. However, due to the sheer size of the analyzed literature, the classification of the number of alternative uses between high, medium, and low should be robust. Even if only 65% of the co-occurrences are the assumed connection, feedstocks with high numbers of alternative uses will still have a much higher number of co-occurrences than the rest. Likewise, feedstocks with low numbers of co-occurrences will not likely be missing enough alternative uses to turn it into a higher classification. In other words, false positives and false negatives will have little impact the classification, so the big picture of the overview is robust.

4.3. Implications for the Expectation of Future Biogas Production in Europe

Our results show that many biogas feedstocks might have multiple alternative uses, and many of these uses have higher value than biogas according to the BVP. This group of feedstocks overlaps with many main feedstocks underlying European biogas expectations. For instance, 8 of the 16 biogas feedstock groups suggested by the EU New Renewable Energy Directive [2] belong to a group with likely a high or medium number of alternative uses. Only for roadside grass and natural grass, we found limited alternatives. If we assume that biomass streams with high and middle alternative uses are not available for biogas production this will have enormous impacts of the biogas potentials. The potential would drop 55%–80% in the two studies [3,5] showing the explicit contributions of each feedstock to the total biogas estimate. In combination with the fact that in present estimates for biogas potentials energy crops are included that are under heavy debate, our research indicates that the existing EU estimate of 1.7 EJ from biogas in 2030 is rather optimistic. In addition, our research suggests other feedstocks with likely low number of alternative uses which might be interesting for biogas production in Table 4.

Table 4. Categorization of European biogas feedstocks expected by current literatures based on likely number of alternative uses as an implication of this research.

Main Feedstocks for Future European Biogas Expected by Current Literatures	Likely Number of Alternative Uses
Manure [2,3,5,7] Straws [2,3,5,7] Organic housewaste [2,3,7] Varied industrial by-products [2,6,7]	High
Varied industrial by-products [2,6,7] Sewage sludge [2,3,7]	Medium
Natural grass [5,7] Roadside grass [3,7]	Low

4.4. Potential Use of the Overview in Research on Biogas Feedstock Competition in a Biobased Economy

Our research can help biogas research and expectations to be framed in the context of a broader transition to the biobased economy. Current research on the topic of biomass competition typically reviews 20–50 applications or aggregated groups of feedstocks [49–52]. By reviewing 109 biogas feedstocks and 217 biomass applications, we identified 1053 individual connections between them in literature which far outnumbers the typical review. In addition, we were able to differentiate groups based on the likely number of alternative applications and whether these uses are higher up the BVP. Thus, our overview can guide research and expectations of future biogas production in taking into account alternative uses of biomass within the biobased economy.

Given the fact that research within specific niches might require a higher level of detail for a few feedstock-application connections, our research can be seen as a filtering tool to identify relevant literature. The machine learning approach was able to collect more than 50,000 articles which contain information of biomass applications and to narrow them down to about 3000 articles that include likely relevant co-occurrences. By reducing the number of articles by an order of magnitude, it turns an unfeasible amount of manual reviewing into a potentially manageable task. When looking at specific feedstocks or applications, the set of articles is reduced even further. Without this machine learning technique, providing a similar overview would be extremely time-consuming or be significantly less comprehensive.

5. Conclusions

In this paper, we provide an overview of biomass streams that can be used for biogas production and their alternative uses. By using the machine learning technique "co-occurrence analysis of terms", the study was able to process a substantial amount of academic literature and identify more than a thousand connections between biogas feedstocks and potential biomass applications.

The overview provides two insights. First, a large share of the biomass streams presently considered in the biogas potential evaluations have many alternative uses in a future biobased economy. In particular, composting-fertilizer-soil amendment and applications related to bioenergy are likely to compete with biogas for biomass feedstocks. This indicates that their contribution to future biogas production is likely to be lower. Second, there are streams not being considered in present policy documents for biogas production although they have the proper characteristics. This shows the advantage of using a value free machine learning process that is able to think out of the box.

Supplementary Materials: The following are available online at http://www.mdpi.com/1996-1073/13/11/2747/s1, Appendix A: List of biogas feedstock terms and variants, Appendix B: List of biomass application terms and variants, Appendix C: List of validated co-occurrences, Appendix D: The co-occurrence matrix and link to the source codes, Appendix E: The list of literature with the co-occurrences.

Author Contributions: Conceptualization, D.L.H., C.D., H.C.M. and S.N.; methodology, D.L.H., C.D., H.C.M. and S.N.; software, C.D.; validation, D.L.H.; formal analysis, D.L.H.; investigation, D.L.H.; data curation, C.D.; writing—original draft preparation, D.L.H.; writing—review and editing, D.L.H., C.D., H.C.M. and S.N.; visualization, D.L.H. and C.D.; supervision, H.C.M. and S.N.; project administration, D.L.H. All authors have read and agreed to the published version of the manuscript.

Funding: This research was funded by NWO- Nederlandse Organisatie voorWetenschappelijkOnderzoek (Dutch Organisation for Scientific Research), Grant/Award Number: 428-15-519 of TKI-Logistiek.

Acknowledgments: We would like to thank people involved in the ADAPNER project for initiating the research theme and their valuable discussions. The ADAPNER project—Adaptive Logistics in Circular Economy is a co-operation between the Faculty of Economics and Business and Faculty of Science and Engineering, University of Groningen. The objective of the ADAPNER project is to study and select viable configurations of value-adding activities related to biomass and biogas in a Circular Economy.

Conflicts of Interest: The authors declare no conflict of interest. The funders had no role in the design of the study; in the collection, analyses, or interpretation of data; in the writing of the manuscript, or in the decision to publish the results.

References

1. Deublein, D.; Seteinhauser, A. *Biogas from Waste and Renewable Resources*; WILEY-VCH Verlag GmbH & Co. KGaA: Weinheim, Germany, 2008; ISBN 9783527318414.
2. European Union. *Directive (EU) 2018/2001 of the European Parliament and of the Council on the Promotion of the Use of Energy from Renewable Sources*; European Parliament: Brussels, Belgium, 2018; Volume 2018.
3. Kampman, B.; Leguijt, C.; Scholten, T.; Tallat-Kelpsaite, J.; Brückman, R.; Maroulis, G.; Lesschen, J.P.; Meesters, K.; Sikirica, N.; Elbersen, B. *Optimal use of Biogas from Waste Streams An. Assessment of the Potential of Biogas from Digestion in the EU Beyond 2020 Digestion in the EU Beyond 2020 Optimal use of Biogas from Waste Streams*; European Commission: Brussels, Belgium, 2016.
4. EU. Com in-depth analysis in support of the commission communication com (2018). In *773 A Clean Planet for all A European Long-Term Strategic Vision for A Prosperous, Modern, Competitive and Table of Contents*; European Commission: Brussels, Belgium, 2018.
5. Meyer, A.K.P.; Ehimen, E.A.; Holm-Nielsen, J.B. Future European biogas: Animal manure, straw and grass potentials for a sustainable European biogas production. *Biomass Bioenergy* **2018**, *111*, 154–164. [CrossRef]
6. Pazera, A.; Slezak, R.; Krzystek, L.; Ledakowicz, S.; Bochmann, G.; Gabauer, W.; Helm, S.; Reitmeier, S.; Marley, L.; Gorga, F.; et al. Biogas in Europe: Food and Beverage (FAB) Waste Potential for Biogas Production. *Energy Fuels* **2015**, *29*, 4011–4021. [CrossRef]
7. Scarlat, N.; Dallemand, J.F.; Fahl, F. Biogas: Developments and perspectives in Europe. *Renew. Energy* **2018**, *129*, 457–472. [CrossRef]
8. Muscat, A.; de Olde, E.M.; de Boer, I.J.M.; Ripoll-Bosch, R. The battle for biomass: A systematic review of food-feed-fuel competition. *Glob. Food Sec.* **2019**, 100330. [CrossRef]
9. Ben Fradj, N.; Jayet, P.A.; Aghajanzadeh-Darzi, P. Competition between food, feed, and (bio)fuel: A supply-side model based assessment at the European scale. *Land Use Policy* **2016**, *52*, 195–205. [CrossRef]
10. Erisman, J.W.; van Grinsven, H.; Leip, A.; Mosier, A.; Bleeker, A. Nitrogen and biofuels; an overview of the current state of knowledge. *Nutr. Cycl. Agroecosystems* **2010**, *86*, 211–223. [CrossRef]
11. Popp, J.; Lakner, Z.; Harangi-Rákos, M.; Fári, M. The effect of bioenergy expansion: Food, energy, and environment. *Renew. Sustain. Energy Rev.* **2014**, *32*, 559–578. [CrossRef]
12. Wu, Y.; Zhao, F.; Liu, S.; Wang, L.; Qiu, L.; Alexandrov, G.; Jothiprakash, V. Bioenergy production and environmental impacts. *Geosci. Lett.* **2018**, *5*. [CrossRef]
13. Paolini, V.; Petracchini, F.; Segreto, M.; Tomassetti, L.; Naja, N.; Cecinato, A. Environmental impact of biogas: A short review of current knowledge. *J. Environ. Sci. Heal. Part. A Toxic Hazard. Subst. Environ. Eng.* **2018**, *53*, 899–906. [CrossRef]
14. Pierie, F.; van Someren, C.E.J.; Benders, R.M.J.; Bekkering, J.; van Gemert, W.J.T.; Moll, H.C. Environmental and energy system analysis of bio-methane production pathways: A comparison between feedstocks and process optimizations. *Appl. Energy* **2015**, *160*, 456–466. [CrossRef]
15. Mohajer, A.; Tremier, A.; Barrington, S.; Teglia, C. Compost mixture influence of interactive physical parameters on microbial kinetics and substrate fractionation. *Waste Manag.* **2010**, *30*, 1464–1471. [CrossRef] [PubMed]
16. Sietske Boschma, D.; Kees, I.; Kwant, W. *Rice Straw Wheat Straw Potential Feedstocks for the Biobased Economy*; NL Agency of Energy and Climate Change: Utrecht, The Netherlands, 2013.
17. Bilo, F.; Pandini, S.; Sartore, L.; Depero, L.E.; Gargiulo, G.; Bonassi, A.; Federici, S.; Bontempi, E. A sustainable bioplastic obtained from rice straw. *J. Clean. Prod.* **2018**, *200*, 357–368. [CrossRef]
18. Maina, S.; Kachrimanidou, V.; Koutinas, A. A roadmap towards a circular and sustainable bioeconomy through waste valorization. *Curr. Opin. Green Sustain. Chem.* **2017**, *8*, 18–23. [CrossRef]
19. Zhang, W.; Qiu, L.; Gong, A.; Cao, Y.; Wang, B. Solid-state Fermentation of Kitchen Waste for Production of Bacillus thuringiensis-based Bio-pesticide. *BioResources* **2013**, *8*, 1124–1135. [CrossRef]
20. Stegmann, P.; Londo, M.; Junginger, M. The circular bioeconomy: Its elements and role in European bioeconomy clusters. *Resour. Conserv. Recycl. X* **2020**, *6*, 100029. [CrossRef]
21. Bosman, R.; Rotmans, J. Transition governance towards a bioeconomy: A comparison of Finland and The Netherlands. *Sustainability* **2016**, *8*, 1017. [CrossRef]
22. Davis, C.B.; Aid, G.; Zhu, B. Secondary Resources in the Bio-Based Economy: A Computer Assisted Survey of Value Pathways in Academic Literature. *Waste Biomass Valorization* **2017**, *8*, 2229–2246. [CrossRef]

23. Shai, S.-S.; Shai, B.-D. *Understanding Machine Learning: From Theory to Algorithms*; Cambridge University Press: New York, NY, USA, 2014; ISBN 978-1-107-05713-5.
24. RStudio | Open Source & Professional Software for data Science Teams-RStudio. Available online: https://rstudio.com/ (accessed on 5 April 2020).
25. ScienceDirect ScienceDirect.com | Science, Health and Medical Journals, Full Text Articles and Books. Available online: https://www.sciencedirect.com/ (accessed on 6 November 2019).
26. ScienceDirect APIs-ScienceDirect | Learn & Support | Elsevier. Available online: https://www.elsevier.com/solutions/sciencedirect/support/api (accessed on 6 November 2019).
27. FAO. *BIOGAS INDUSTRIAL User Manual Bioenergy and Food Security Rapid Appraisal (BEFS RA)*; Food and Agriculture Organization of the United Nations: Rome, Italy, 2017.
28. FAO. Energy end-use options | Energy | Food and Agriculture Organization of the United Nations. Available online: http://www.fao.org/energy/bioenergy/befs/assessment/befs-ra/energy-end-use/en/ (accessed on 18 April 2020).
29. KTBL. *European Feedstock Atlas-Feedstock List*; KTBL: Darmstadt, Germany, 2010.
30. KTBL. Feedstock Atlas. Available online: https://daten.ktbl.de/euagrobiogasbasis/startSeite.do;jsessionid=0B98622076229B9E15308366CEE1623B (accessed on 18 April 2020).
31. Feedipedia Feedipedia: An on-line encyclopedia of animal feeds | Feedipedia. Available online: https://www.feedipedia.org/ (accessed on 18 April 2020).
32. Feedbase. FeedBase.com-Economic and technical feed information. Available online: http://www.feedbase.com/index.php?Lang=E (accessed on 18 April 2020).
33. Luske, B.; Blonk, H. *Milieueffecten van Dierlijke Bijproducten Blonk Milieuadvies Werken aan Duurzaamheid*; Blonk Milieu Advies: Gouda, The Netherlands, 2009.
34. Quik, J.T.K.; Mesman, M.; van der Grinten, E. *Assessing Sustainability of Residual Biomass Applications Finding the Optimal Solution for A Circular Economy*; Rijksinstituut voor Volksgezondheid en Milieu RIVM: Bildhofen, The Netherlands, 2016.
35. Bos-Brouwers, H.; Langelaan, B.; Sanders, J. *Chances for Biomass*; Wageningen University & Research: Wageningen, The Netherlands, 2012.
36. Wang, S.J.; Huang, S.H.; Wong, D.S.H.; Jang, S.S. Novel plant-wide process design of dichlorohydrin production by glycerol hydrochlorination. *J. Taiwan Inst. Chem. Eng.* **2017**, *73*, 50–61. [CrossRef]
37. Okoye, P.U.; Abdullah, A.Z.; Hameed, B.H. Synthesis of oxygenated fuel additives via glycerol esterification with acetic acid over bio-derived carbon catalyst. *Fuel* **2017**, *209*, 538–544. [CrossRef]
38. Zhou, H.; Long, Y.Q.; Meng, A.H.; Li, Q.H.; Zhang, Y.G. Interactions of three municipal solid waste components during co-pyrolysis. *J. Anal. Appl. Pyrolysis* **2015**, *111*, 265–271. [CrossRef]
39. Rafiee, A.; Gordi, E.; Lu, W.; Miyata, Y.; Shabani, H.; Mortezazadeh, S.; Hoseini, M. The impact of various festivals and events on recycling potential of municipal solid waste in Tehran, Iran. *J. Clean. Prod.* **2018**, *183*, 77–86. [CrossRef]
40. Markidis, I.; Komilis, D.; Tsagas, F.; Petalas, A. A fractional factorial field experiment to study the decomposition of municipal solid wastes stored in wrapped bales. *J. Environ. Manage.* **2013**, *115*, 32–41. [CrossRef] [PubMed]
41. Yao, R.S.; Zhang, P.; Wang, H.; Deng, S.S.; Zhu, H.X. One-step fermentation of pretreated rice straw producing microbial oil by a novel strain of Mortierella elongata PFY. *Bioresour. Technol.* **2012**, *124*, 512–515. [CrossRef] [PubMed]
42. Yang, S.; Li, Q.; Gao, Y.; Zheng, L.; Liu, Z. Biodiesel production from swine manure via housefly larvae (Musca domestica L.). *Renew. Energy* **2014**, *66*, 222–227. [CrossRef]
43. Lu, J.; Watson, J.; Zeng, J.; Li, H.; Zhu, Z.; Wang, M.; Zhang, Y.; Liu, Z. Biocrude production and heavy metal migration during hydrothermal liquefaction of swine manure. *Process. Saf. Environ. Prot.* **2018**, *115*, 108–115. [CrossRef]
44. Theegala, C.S.; Midgett, J.S. Hydrothermal liquefaction of separated dairy manure for production of bio-oils with simultaneous waste treatment. *Bioresour. Technol.* **2012**, *107*, 456–463. [CrossRef]
45. Li, Q.; Zheng, L.; Cai, H.; Garza, E.; Yu, Z.; Zhou, S. From organic waste to biodiesel: Black soldier fly, Hermetia illucens, makes it feasible. *Fuel* **2011**, *90*, 1545–1548. [CrossRef]
46. Contreras-Ramos, S.M.; Escamilla-Silva, E.M.; Dendooven, L. Vermicomposting of biosolids with cow manure and oat straw. *Biol. Fertil. Soils* **2005**, *41*, 190–198. [CrossRef]

47. Mandl, M.G. Status of green biorefining in Europe. *Biofuels Bioprod. Biorefining* **2010**, *6*, 268–274. [CrossRef]
48. Xiu, S.; Shahbazi, A. Development of Green Biorefinery for Biomass Utilization: A Review. *Trends Renew. Energy* **2015**, *1*, 4–15. [CrossRef]
49. Gerssen-Gondelach, S.J.; Saygin, D.; Wicke, B.; Patel, M.K.; Faaij, A.P.C. Competing uses of biomass: Assessment and comparison of the performance of bio-based heat, power, fuels and materials. *Renew. Sustain. Energy Rev.* **2014**, *40*, 964–998. [CrossRef]
50. Pavlenko, N.; El Takriti, S.; Malins, C.; Searle, S. *Beyond the Biofrontier: Balancing Competing Uses for the Biomass Resource*; International Council on Clean Transportation: Washington, USA, 2016.
51. Philippidis, G.; Bartelings, H.; Helming, J.; M'barek, R.; Smeets, E.; van Meijl, H. Levelling the playing field for EU biomass usage. *Econ. Syst. Res.* **2019**, *31*, 158–177. [CrossRef]
52. Cherubini, F. The biorefinery concept: Using biomass instead of oil for producing energy and chemicals. *Energy Convers. Manag.* **2010**, *51*, 1412–1421. [CrossRef]

Article

Syngas Production, Clean-Up and Wastewater Management in a Demo-Scale Fixed-Bed Updraft Biomass Gasification Unit

Gabriele Calì [1,*], Paolo Deiana [2], Claudia Bassano [2], Simone Meloni [1], Enrico Maggio [1], Michele Mascia [3] and Alberto Pettinau [1]

[1] Sotacarbo S.p.A., Grande Miniera di Serbariu, 09013 Carbonia, Italy; simone.meloni@sotacarbo.it (S.M.); enrico.maggio@sotacarbo.it (E.M.); alberto.pettinau@sotacarbo.it (A.P.)

[2] ENEA, Italian Agency for New Technologies, Energy and Sustainable Economic Development, Via Anguillarese 301, 00123 Roma, Italy; paolo.deiana@enea.it (P.D.); claudia.bassano@enea.it (C.B.)

[3] Dipartimento di Ingegneria Meccanica, Chimica e dei Materiali, Università degli Studi di Cagliari, Piazza D'Armi, 09123 Cagliari, Italy; michele.mascia@unica.it

* Correspondence: gabriele.cali@sotacarbo.it

Received: 20 April 2020; Accepted: 17 May 2020; Published: 20 May 2020

Abstract: This paper presents the experimental development at demonstration scale of an integrated gasification system fed with wood chips. The unit is based on a fixed-bed, updraft and air-blown gasifier—with a nominal capacity of 5 MW_{th}—equipped with a wet scrubber for syngas clean-up and an integrated chemical and physical wastewater management system. Gasification performance, syngas composition and temperature profile are presented for the optimal operating conditions and with reference to two kinds of biomass used as primary fuels, i.e., stone pine and eucalyptus from local forests (combined heat and power generation from this kind of fuel represents a good opportunity to exploit distributed generation systems that can be part of a new energy paradigm in the framework of the circular economy). The gasification unit is characterised by a high efficiency (about 79–80%) and an operation stability during each test. Particular attention has been paid to the optimisation of an integrated double stage wastewater management system—which includes an oil skimmer and an activated carbon adsorption filter—designed to minimise both liquid residues and water make-up. The possibility to recycle part of the separated oil and used activated carbon to the gasifier has been also evaluated.

Keywords: biomass gasification; demonstration-scale plant; syngas; circular economy; wastewater management; activated carbon adsorption

---

1. Introduction

The increasing attention towards climate change and greenhouse gas emissions makes the exploitation of renewable energy sources one of the key pathways for sustainable power generation. It is expected to involve a significant reduction (some 8 Gt/yr by 2050, with a share of 32% among the other low carbon approaches) of CO_2 emissions in the power generation sector, according to the most recent assessment by the International Energy Agency [1]. The same target has been formally assumed by the European Union (EU) with the publication, in December 2018, of the revised "renewable energy directive" (2018/2001/EU), which aims at keeping the EU a global leader in renewable energy and to meet the commitments under the Paris Agreement [2,3]. However, the diffusion of intermittent sources (i.e., wind and solar) makes grid regulation increasingly challenging since ever changing electrical loads must be balanced with ever changing, non-programmable generation [4]. On the contrary, bioenergy can be considered a key option to mitigate greenhouse gas emissions, replace fossil fuels, and—considering its programmable exploitation—ensure a more secure and sustainable energy system [5]. Additionally, the use of waste biomass (e.g., agro-industrial, municipal and forestry residues) [6] is even more interesting since it allows production of almost CO_2-free energy as an alternative to landfill or inefficient biological processes.

Biomass gasification is an appealing thermochemical conversion process that allows the production of a synthesis gas (syngas) that can be used for power generation in internal combustion engines [7,8] or for other applications, such as liquid fuels production [9].

Gasification technologies for small- and medium-scale combined heat and power (CHP) generation from waste biomass have been significantly developed in the last decades [10]. Most of the attention is focused on fixed-bed downdraft processes, typically available for a capacity between 200 and 700 kW_{th} [11–13], but few studies also consider bubbling [14,15] or circulating fluidised-bed [16] gasification processes, sometimes promoted with specific catalysts [17] or integrated with hydrothermal carbonisation to treat high-moisture biomass [18]. With respect to these technologies, fixed-bed updraft gasification allows a better conversion efficiency (thanks to the countercurrent heat exchanges) [4] and it is typically characterised by simple construction, easy operation, fuel flexibility in terms of type (biomass, coal, wastes, etc.), particle size (5 to 100 mm) and moisture content (up to 60%) [19], but also involves a relatively high production of pyrolysis liquids (i.e., oils and tar) [4]. Such a technology is feasible for applications in the order of a few thermal megawatts.

In fixed-bed updraft gasification reactors, the solid primary fuel is loaded from the top of the reactor and supported by a metallic grate by which the gasification agents (air or oxygen and possibly steam, depending on the specific process) are injected from the bottom, allowing a countercurrent. As fuel flows downwards, it is heated by the hot raw gas that moves upwards, coming from the gasification and combustion zones [20,21], whereas the gasification agents are preheated by cooling the bottom ash [20]. The units developed for small- and medium-scale applications typically operate at atmospheric pressure using air (instead of oxygen, used for the industrial-scale processes) and possibly steam as gasification agents. A number of theoretical studies on these kinds of processes are currently available in the scientific literature, mainly focused on the development of thermodynamic models (in particular by minimising Gibbs free energy) [22,23] or on computational fluid dynamics (CFD) process simulation [24,25], in some cases with the experimental model validation in pilot units [26,27]. Several studies are available on pilot-scale experimental development of the process for waste biomass gasification for power generation and biochar [7,28]. To the best of the authors' knowledge, just a few studies are specifically focused on air-blown fixed-bed updraft technology demonstration at commercial scale. Only Lei et al. [29] recently published an experimental research based on a batch feeding updraft gasifier designed to treat 2 tons per day of rural solid waste in China.

A major issue in biomass gasification plants is the managing of tar, which is the complex mixture of condensable hydrocarbons generated during gasification of such volatile-rich matter as biomasses [30]. Tar is constituted by single-ring to 5-ring aromatic compounds plus other oxygen-containing hydrocarbons and complex polyaromatic hydrocarbons (PAHs), phenolic compounds (cresols, xylenols, etc.), and monocyclic aromatic hydrocarbons such as benzene, toluene and xylene [31,32]. Tar formation is still considered to be the main technological barrier of this kind of technology [22], since it causes several environmental and industrial problems: heavy tars may condense on cooler surfaces downstream leading to blockage of filters and fuel lines [31], it can block valves and clog fuel injectors of engines [33] and, if released in the environment, it can have harmful effects, since its compounds are toxic and potentially carcinogens [34,35]. Tar issues are particularly relevant for updraft gasifiers—with a loading in raw syngas of about 50 g/Nm^3 [36]—since the countercurrent heat exchanges involve relatively low temperatures in the upper part of the fuel bed, promoting tar formation and limiting its decomposition into lighter compounds. Therefore, its management is of great relevance to make gasification a feasible option for power generation from biomass. Tar management can be basically achieved through two strategies: reduction of tar formation inside the gasifier, with the so-called primary methods, or separation after the gasification process, with secondary methods [31]. Primary methods include optimal design of the gasifier, the optimisation of process parameters, and possibly (depending on the specific technology) the use of suitable additives or catalysts. Secondary methods comprise thermal or catalytic cracking, or mechanical methods such as the use of cyclones and electrostatic filters [37], as well as the use of wet scrubbers. The latter, in particular, is an effective and reliable process for tar removal from syngas. However, the process may generate a high amount of wastewater. The toxicity of many compounds of tar makes the wastewater treatment with biological processes quite difficult. Therefore, chemical and physical treatment must be used, such as advanced oxidation processes, or precipitation with Fe or Al salts [23]. Adsorption on activated carbon is also an effective technology to remove tar from wastewater, although regeneration or disposal of the exhaust sorbent is a drawback of the process [38].

This paper aims to establish the baseline performance of a 5 MW$_{th}$ (1.3 m internal diameter) demonstration-scale updraft gasifier, operating since 2014 at the Sotacarbo Research Centre in Sardinia, Italy [39], and tested for some 1500 h with different kinds of coal and biomass. In particular, the experimental results here reported are focused on the gasification of two kinds of local waste biomass, with the aim to assess operating conditions, syngas composition and properties and the whole plant performance. In addition, the syngas cleaning and wastewater treatment process performance is also evaluated, on the basis of a novel configuration of a tar management system optimised to minimise water consumption and sludge disposal and recirculate part of the separated tar and exhaust activated carbon to the gasification unit (thus improving the efficiency of the whole project).

2. Materials and Method

The analysis presented here summarises the key results of an experimental campaign carried out in order to improve the knowledge of the gasifier's operation in different operating phases (i.e., start-up, steady-state and shut down) and to evaluate the performance of the integration with tar management and wastewater treatment. Moreover, the final scope is to provide useful data so as to improve the system efficiency to make electricity generation suitable by mean of an internal combustion engine fed by clean syngas.

2.1. The Sotacarbo Demonstration-Scale Gasification Unit

The nominal 5 MW$_{th}$ demonstration-scale plant (Figure 1) of the Sotacarbo platform is mainly composed of a gasification section, a wet scrubber for syngas clean-up, a wastewater treatment system and a flare for final syngas combustion, according to the simplified scheme shown in Figure 2.

Figure 1. The Sotacarbo demonstration-scale unit.

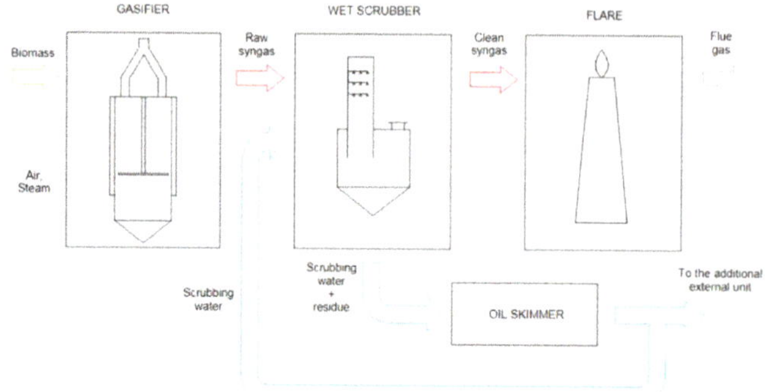

Figure 2. Simplified scheme of the demonstration-scale unit.

2.1.1. Gasification Reactor

The key section of the experimental unit is the gasification reactor. It is based on the historical Wellman-Galusha fixed-bed updraft technology, which has a commercial history in municipal and industrial applications dating back at least 80 years [40]. It was further developed in the United States by Hamilton Maurer International, Inc. (HMI, Houston, TX, USA) for power generation from coal during more than 10,000 h between 1981 and 1985 [41]. It was converted for biomass (wood chips) gasification by Ansaldo Energia (former Ansaldo Ricerche), tested in a 1.3 m internal diameter gasifier, and put in operation between 1999 and 2001 [4].

Biomass is loaded by means of an automatic redler conveyor and it is introduced in the top of the reactor through four different injection points. The fuel bed is supported by an eccentric grate that allows continuous ash discharge. The grate design (with different coplanar plates) has been optimised to allow the reactor to operate in continuous mode, avoiding blockages of the ash extraction system. It is equipped with a robust driving system specifically designed to optimise its operation. The gasification agents (air and/or steam) are injected from the bottom of the reactor through the grate at about atmospheric pressure (with a small overpressure just to win the pressure drops through the whole process). As fuel flows downwards, it is heated by the hot raw gas that moves upwards, coming from the grate [20], so that the following processes take place: fuel drying, devolatilisation, pyrolysis, gasification and partial combustion [38].

The gasifier is also equipped with a cooled stirrer composed of a vertical shaft, with internal water recirculation, which stirs and uniforms the fuel bed in order to maximise the process performance. This device can translate vertically and perform both clockwise and counterclockwise rotation. Moreover, the reactor wall is constantly cooled thanks to a water jacket that allows a slight thermal dissipation. Steam generated in the jacket during plant operations, reaches by natural circulation the upper steam drum, downstream connected to a forced air-cooled condenser.

Due to the experimental nature of the unit, characterised by frequent start-up and shut-down phases, the ignition phase is performed by six infrared ceramic irradiators (instead of the conventional burners) placed above the bottom of the reactor, which is also equipped with several thermo-couples located on different levels, in order to monitor the internal temperature profile.

2.1.2. Wet Scrubber

Gas cleaning is an essential component of any biomass gasification plant to meet the specifications of the syngas end user. Internal combustion engines for power generation require limits of about 30 mg/Nm^3 for particulate and 100 mg/Nm^3 for tar [36,42]. Conventionally, gas cleaning is performed by means of different systems, such as wet or wet-dry scrubbing (the former being the most common in this kind of applications) or hot gas conditioning.

In this specific case, raw syngas from the gasification unit is sent to a wet scrubbing system (Figure 3) by means of an insulated pipe (to reduce tar condensation), installed with a slight slope to allow condensates to move to the scrubber by gravity. The scrubber operates in co-current mode, in order to remove tar and dust and to prevent backfire. The scrubbing tower is provided with a dehydration section for the separation of the micro drops of water dragged in the purified syngas. Heavy non soluble tar (C20–C40 and more), together with inert matter and un-reacted dust separated from syngas, is collected in the conic bottom of the scrubber and removed through a screw pump [43]. On the other hand, water with light tar (C10–C20) moves to the oil skimmer section.

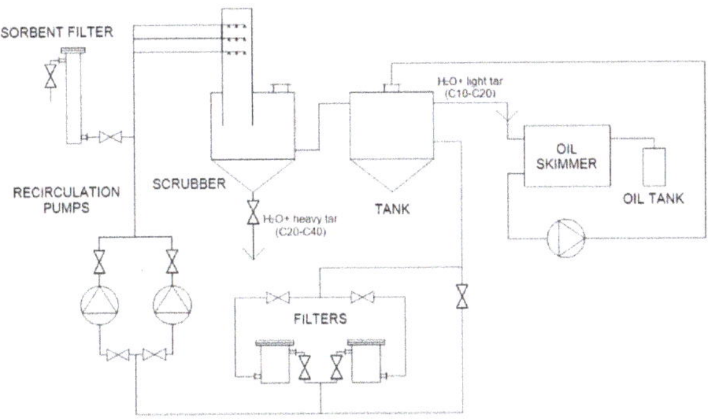

Figure 3. Water recirculation system.

2.1.3. Wastewater Treatment

The wastewater leaving the scrubber (and temporarily stored in a 6 m^3 tank that allows a thermal flywheel effect) is loaded with tar, dust and various contaminants and must be treated before being recirculated. As mentioned, the integrated wastewater treatment system (schematically represented in Figure 3) has been specifically designed to minimise freshwater consumption and sludge disposal and to recirculate part of the separated tar and exhaust activated carbon to the gasification unit (thus recovering its energy).

Firstly, wastewater from the scrubber is pre-treated by means of an oil skimmer (specifically designed and assembled using only commercial components to contain capital cost, in view of commercial applications), with the aim to continuously separate insoluble oils and tar from the surface of the liquid phase (thanks to the low specific weight of the compounds and their affinity with the materials) and potentially to recirculate a part of the pre-treated water directly to the scrubbing system. About 8 m^3/h of wastewater is collected from the tank, filtered and recirculated to spray nozzles of the wet scrubber through proper pumps. On the other hand, about 0.4 m^3/h of light tar mixed with water (about 5% of the scrubber wastewater recirculation flow) is collected from the free surface of the tank and delivered to the oil skimmer. An additional external unit provides further treatment in a proper chemical–physical unit, equipped with different sections for chemical reagent dosing, destabilisation, flocculation, sedimentation and sludge purging. Treated water is then collected in an external tank before disposal.

Among the available technologies (i.e., wet oxidation, adsorption on active carbon and/or carbon-rich ashes from gasification, as well as chemical, physical and biological treatment [43], the latter not feasible due to the toxicity of the components to be treated [23]), a chemical–physical treatment unit has been selected.

Physical treatment as ultraviolet light-induced wet oxidation or adsorption on various coke sorbents was suggested to treat this type of wastewater. Moreover, chemical precipitation using various salts of Fe and Al can promote the formation of flocs, then reducing the concentration of colloidal and particulate matter in the wastewater [23]. Finally, adsorption on activated carbon is demonstrated to be an effective technology to remove organic pollutants from wastewaters. The main benefits of using activated carbon are the large surface area, the economic viability and the easy operational procedures [44], whereas the disadvantages are a difficult and expensive regeneration process that can limit its application, considering also that the spent carbon generally involves a serious disposal problem [38].

In this specific case, the different commercial diluted solutions dosed in this unit are based on the following chemical reagents: hydrochloric acid for pH neutralisation, ferric chloride for destabilisation and flocculation [45], calcium hydroxide as a coagulant, polyelectrolytes as a thickener. The reactions take place in different phases, starting with destabilisation and coagulation and ending with flocculation, and are carried out in two separate tanks, equipped with mechanical stirrers set at different speeds (the first fast and the second slow). The separation between purified water and sludge takes place in a tank equipped with lamellar septa and a single-screw mud pump for sludge extraction.

2.1.4. Auxiliaries

The unit is also equipped with different auxiliary systems to allow the experimental runs: fuel temporary storage and charging system, dust extraction and filtration system, process air and steam production and adduction systems, and LPG storage and adduction system (to support syngas combustion by the flare). For safety reasons, such pipeline inertisation or emergency shutdown of the plant, require a nitrogen storage and adduction system; it comprises a vertical cryogenic vessel for liquid nitrogen and a system of vaporisers.

The plant includes industrial data acquisition and control equipment in order to continuously monitor the main process parameters, with particular reference to temperatures, pressures, and flow rates of the gasification agents. Particular attention is dedicated to monitor the internal temperature profiles of the gasifier, crucial for correct operations and for reliable data processing.

Composition of raw syngas from the gasifier, clean syngas downstream of the scrubber, exhausts from flare and vents are monitored by four different gas sampling lines. A real time multi module industrial analysis system and a portable micro gas chromatograph (GC) are dedicated to continuously monitor the syngas composition. The first provides a quick online measure of H_2, CO, CO_2, CH_4, H_2S and O_2. It consists of several gas analysers, which provide: H_2 concentration measured by thermal conductivity by means of a CALDOS 25 module (within a range between 0% and 100% by volume); CO, CO_2 and CH_4 concentrations measured by an infrared URAS26 module (within the following ranges: 0–30% for CO, 0–25% for CO_2, and 0–5% for CH_4, all by volume); H_2S concentration measured by an ultraviolet Limas 11 module (between 0% and 2% by volume); and O_2 concentration measured by a paramagnetic Magnos 206 module (between 0% and 25% by volume). Moreover, a micro GC (Agilent 2000, Santa Clara, CA, USA) analyses H_2, CO, CO_2, CH_4, H_2S, and O_2 in the syngas, as well as N_2, COS, C_2H_6 and C_3H_8. The plant is also equipped with a sampling line that allows measuring the amount of tar at the exit of the gasifier. To this aim, syngas is collected downwards of the reactor and sent—through a hot line (200 °C)—to three ice cooled traps to allows tar condensation and sampling. Samples are then analysed in the laboratory, typically in terms of weight, calorific value and composition.

2.2. Primary Fuels

The experimental campaign presented here has been carried out with stone pine (*Pinus pinea*) and eucalyptus (*Eucaliptus camaldulensis*) wood chips from local wood management. For each biomass, two different samples have been delivered and analysed, with slight differences mainly in terms of moisture content.

Table 1 shows, for each sample, proximate and ultimate analysis, as well as lower heating value (LHV) and bulk density. The analyses have been carried out in the Sotacarbo laboratories according to the international standards. In particular, proximate analysis has been performed by a LECO TGA-701 thermogravimetric analyser; ultimate analysis has been carried out on a LECO Truspec CHN/S analyser;

finally, the energy content of the sample (higher heating value) has been measured using an adiabatic oxygen bomb calorimeter by a LECO AC-500 calorimeter, according to the ISO 1928:1995 standard, and then converted into LHV considering the moisture content.

Table 1. Characterisation of fuel samples, as received.

Parameter	Stone Pine (1)	Stone Pine (2)	Eucalyptus (1)	Eucalyptus (2)	Standard
Proximate analysis (% by weight)					
Fixed carbon	22.33	21.09	15.37	15.67	By difference
Volatiles	61.87	66.91	56.09	57.41	ASTM D 5142-04
Moisture	11.93	9.25	26.86	25.76	ASTM D 5142-04
Ash	2.87	2.75	1.68	1.16	ASTM D 5142-04
Ultimate analysis (% by weight)					
Total carbon	50.72	50.88	49.62	49.75	ASTM D 5373-02
Hydrogen	6.32	6.71	6.17	6.54	ASTM D 5373-02
Nitrogen	0.63	0.50	0.48	0.35	ASTM D 5373-02
Oxygen	27.53	29.91	15.19	16.54	By difference
Moisture	11.93	9.25	26.86	25.76	ASTM D 5142-04
Ash	2.87	2.75	1.68	1.16	ASTM D 5142-04
Other properties					
LHV (MJ/kg)	16.07	16.08	13.12	13.34	
Bulk density (kg/dm^3)	0.20	0.22	0.25	0.26	

2.3. Experimental Procedures

Basically, each run is organised in four different phases: plant preparation, start-up (which typically lasts for about four hours), steady-state operation and plant shut-down (that requires about five hours). Before each run, all the auxiliary systems are started up and their functionality is verified, in order to assure the right operation of the whole plant. In parallel, some instruments are calibrated to avoid significant errors in the measurement of the key parameters. As soon as the auxiliary operation is verified, the start-up process can begin: primary fuel is heated by means of the six ceramic infrared irradiators, until fuel locally reaches the temperature of 750–800 °C (typically this phase takes some 45 min, with 15 extra min to complete fuel heating); then air is injected through the ceramic irradiators to start the fuel ignition; in this phase, gas is vented in the atmosphere. When the ignition of the fuel bed is confirmed (it needs just few additional minutes), the ceramic irradiators are turned off, the air flow through the irradiators is stopped, a sub-stoichiometric air flow is injected through the grate and gas is sent to the flare and burnt; this can be consider the beginning of the operation phase [39]. In general, during the operation phase, a steady-state regime is kept for several hours (through a continuous feeding of the primary fuel selected for the specific run) or pre-determined operating procedures are followed, on the basis of the specific aims of each run. The shut-down phase of the plant begins at the end of the experimental test, interrupting the fuel loading and decreasing air flow to cool the reactor down.

The runs are operated with a primary fuel (as-received) characterised by a particle size between 10 and 50 mm (with an amount of fines within 5% by weight), maintaining the main operating parameters constant (in particular air flow) and regulating fuel loading to achieve an almost constant fuel bed level (at about 1800 mm), with a maximum fluctuation of ±30 mm.

During the experimental tests, wastewater samples are collected from different sections of the circuit with a pre-determined frequency: from the bottom of the scrubber tank (every 3 h), from a floating skimmer (1 h), in the decantation tank (1 h), from the rotating oil skimmer (3 h), and downstream of the activated carbon filter (5 to 10 min). Samples are then analysed in the Sotacarbo labs by measuring pH (by means of a portable pH-meter Medidor PH BASIC 20 and a Yokogawa online pH-meter) and suspended and dissolved solids.

3. Results and Discussion

Gasification results that were collected, for each fuel sample, during about 80 h of continuous steady-state operation have been processed in order to determine plant operating conditions and performance. In parallel, wastewater has been collected and analysed. With respect to the previously published data [39], referred to a preliminary biomass gasification experimental campaign, the results here reported come from a specific optimisation of the operating procedures for this kind of fuel.

3.1. Gasification Performance

The experimental tests have been arranged regulating the process operating parameters (mainly fuel and air injection) to optimise syngas composition and maximise its lower heating value as well as process stability. Table 2 reports a summary of the key operating parameters (including the equivalence ratio, ER, defined as the ratio between the oxygen actually injected as gasification agent and the oxygen theoretically required for the stoichiometric fuel combustion) and the subsequent syngas composition, intended as the average value during the steady-state operation. In particular, the key index assumed as a measure of the global performance of the gasification process is the so-called cold gas efficiency (η_{CG}), conventionally defined on the basis of the first law of thermodynamics as the ratio between the chemical energy of raw syngas (calculated as the product of syngas mass flow and its lower heating value) and the chemical energy of primary fuel:

$$\eta_{CG} = \frac{m_{syn} \cdot LHV_{syn}}{m_{fuel} \cdot LHV_{fuel}} \tag{1}$$

where m is the mass flow rate (in kg/s) and LHV is the lower heating value (in MJ/kg) of syngas and fuel, indicated by the subscripts *syn* and *fuel*, respectively.

First of all, it can be noticed that raw syngas from eucalyptus chips gasification is characterised by a slightly higher CO_2 concentration and a lower CO concentration than those measured during the stone pine gasification tests. This is one of the effects of the higher moisture content of eucalyptus biomass, which promotes the water–gas shift reaction. In parallel, it can be noticed that eucalyptus runs have been performed with a lower specific air flow rate (about 1.9 kg of air per kilogram of biomass, compared with 2.2 kg/kg for stone pine, both previously determined as the optimum values for the considered fuels), which leads to a slightly higher heating value. Cold gas efficiency is in the order of 79–80% for all the runs. Moreover, with respect to eucalyptus gasification, the higher equivalence ratio used for stone pine gasification led to higher temperatures in the combustion zone and in the freeboard with, as a consequence, a significantly lower tar production.

Figure 4 shows the trend with time of H_2, CO_2, CO and CH_4 concentrations in the syngas from one of the runs with stone pine chips (similar results have been obtained during the other three runs). Syngas composition is quite constant during the whole steady-state operation. It can be also noticed that the gasification conditions have been reached after about 3 h from the beginning of the start-up procedure, with an increasing of CO and H_2 concentration, whereas the steady state is reached after about 10 h with the stabilisation of syngas composition.

One of the key aspects for the evaluation of a fixed-bed updraft gasifier operation is represented by the temperature profile into the reactor, shown in Figure 5—with reference to wall temperatures—for the same pine-fueled run. Excluding the start-up phase, when the whole reactor is at almost ambient temperature, the profile shows good stability during the whole run. The trend reveals the characteristic tendency of fixed bed updraft gasifiers with the maximum temperature in the combustion zone [39]. And, thanks to the presence of the water jacket, the inner temperature in the freeboard (above about 1000 mm from the bottom of the reactor) is typically lower than 200 °C and around 100 °C at the outlet of the gasifier. This involves the condensation of the heaviest tar compounds, which remain trapped in the reactor in a sort of natural recirculation.

Table 2. Operating parameters and syngas properties.

Parameter	Stone Pine (1)	Stone Pine (2)	Eucalyptus (1)	Eucalyptus (2)
Operating parameters				
Fuel loading (kg/h)	280	275	345	340
Air flow (kg/h)	630	620	655	650
Equivalence ratio (%)	33.02	32.85	26.48	26.36
Fuel bed level (mm)	1800	1800	1800	1800
Maximum temperature (°C)	870	880	830	824
Freeboard temperature (°C)	380	376	300	310
Syngas composition (molar fraction, dry basis)				
CO	0.29	0.29	0.27	0.28
CO_2	0.05	0.05	0.06	0.06
H_2	0.12	0.12	0.12	0.12
CH_4	0.02	0.02	0.02	0.02
H_2S	0.00	0.00	0.00	0.00
N_2	0.52	0.52	0.53	0.52
Syngas properties and process performance				
Syngas mass flow (kg/h)	860	845	918	910
Syngas mass flow (kg/h, dry)	783	772	839	832
Lower heating val. (MJ/kg, dry)	4.58	4.58	4.29	4.29
Specific heat (kJ/(kg·K))	1.11	1.11	1.12	1.12
Cold gas efficiency	0.797	0.800	0.795	0.787
Byproducts				
Tar production (kg/h)	50	45	80	92
Ash production (kg/h)	5.5	7.0	5.0	5.4

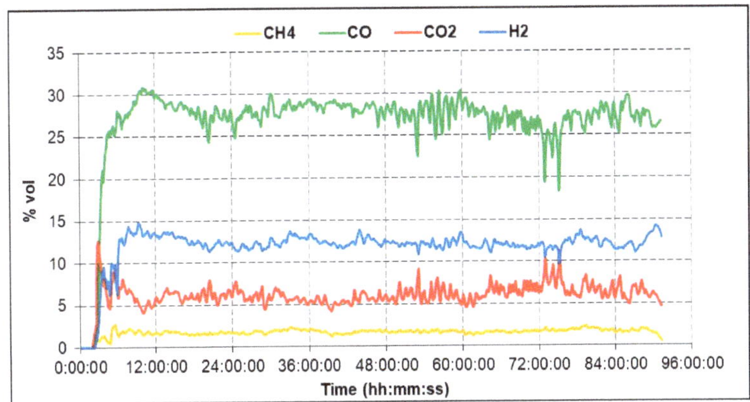

Figure 4. Raw syngas composition trend for stone pine gasification.

If the cold gas efficiency is the key performance indicator for the gasification process, a general performance on the whole plant (excluding the power generation system, not considered in this study) can be assessed on the basis of a carbon balance, based on the simplified scheme reported in Figure 6.

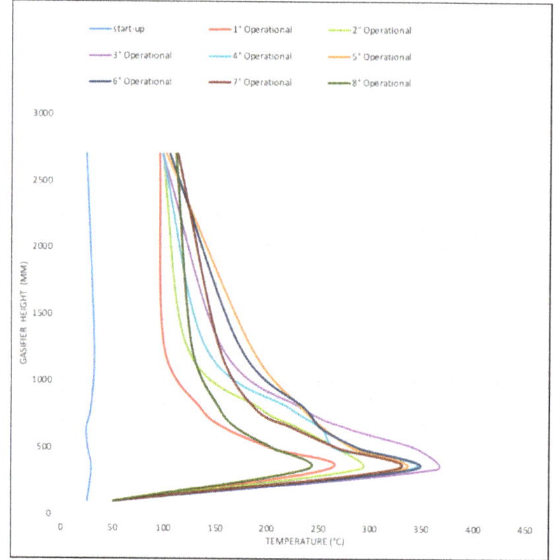

Figure 5. Wall temperature profiles for stone pine gasification.

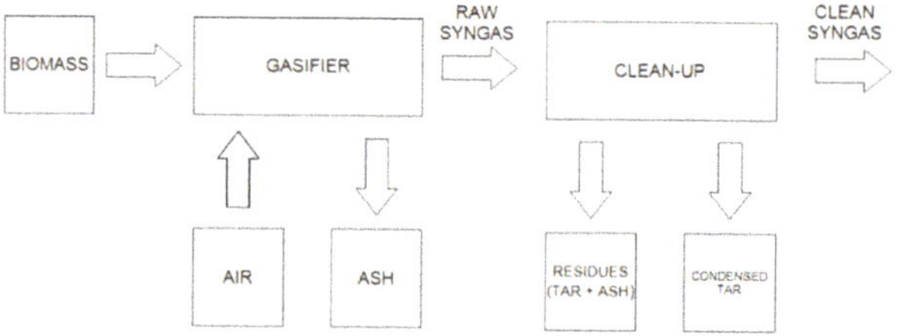

Figure 6. Process scheme for carbon balance assessment.

Globally, carbon balance can be easily represented by the following equation:

$$C_{in} = C_{out} + C_{acc} \qquad (2)$$

where C_{in} is carbon that enters the plant through the primary fuel, C_{out} is carbon that leaves the plant and C_{acc} represents carbon accumulation within the process. In particular, C_{out} includes carbon contained in bottom ash (which is negligible, indicating a very efficient conversion during the gasification process), in the scrubber residues (some 50 kg/h of heavy tars and dust discharged from the bottom of the scrubber), in light condensed tar (about 10 kg/h separated from the free surface of the scrubber and discharged by the oil skimmer), and in clean syngas as CO, CO_2 and CH_4. The residues analyses are summarised in Table 3. On the other hand, C_{acc} does not include any contribution from the gasification unit (the process works at steady state and the fuel level into the gasifier is kept almost constant during the whole run), whereas some accumulation occurs in the wet scrubber (carbon residues in the washing water) and in the filters of the water recirculation system (about 7 kg/h). Due to these accumulations, and in order not to saturate the washing water and keep the water levels constant in the tanks, a make-up of fresh water of about 60 kg/h is required.

Table 3. Residues characterisation (% by weight).

Parameter	Bottom Ash	Liquid Residues	Condensed Tar
Fixed carbon	0.14	n.a.	n.a.
Volatiles	18.66	0.00	0.00
Moisture	0.13	30.00	20.00
Tars (C10–C40)		68.00	80.00
Minerals	81.08	2.00	0.00
Lower heating value (MJ/kg, dry basis)	n.a.	26.00	22.00

3.2. Wastewater Treatment Optimisation and Performance

The efficiency of the implementation of an oil skimmer in the scrubber loop system is evaluated in terms of pH, and removal of total suspended solids (TSS). Wastewater from the scrubber tank is characterised by pH values from 3.5 to 4.5, with TSS of about 200 mg/dm^3 (measured by filtration and drying of the collected samples).

During the first gasification campaigns (before the installation of the oil skimmer), the content of tar and inorganic compounds in clean-up water tended to increase, while pH decreased down to a pseudo steady-state value of about 3.5. Lab scale titration experiments with 1.2 M HCl showed a high buffer capacity of the clean-up water around this value. This behaviour has been confirmed also after CO_2 removal from the samples by insufflation of air: the separation between liquid and solid phases (the latter constituted by the heavy tar deposited on the bottom of the scrubber) was observed. Light tar is partially separated by the bubbles rising in the liquid phase in the scrubber, and forms a layer above the free water surface. The dissolved fraction of tar is almost completely recirculated in the liquid stream and tends to accumulate, increasing the organic load with time. The high content of pollutants in the clean-up water reduces the effectiveness of scrubbing and may lead to fouling of surfaces and blocking of the recirculation pumps. Different strategies have been implemented or planned to remove organics from the clean-up water with in-line processes, to increase the effectiveness of the scrubbing and reduce water make-up. Furthermore, this complete in-line treatment system makes it possible not to use the chemical separation process, with a decrease in the investment and operation costs. In particular, the residues, consisting mainly of tar with a high specific weight C20–C40, are expelled from the bottom of the scrubber tank where washing takes place. The tar with low specific weight (C10–C20), floating on the top layer of the tank, is taken from the oil skimmer, concentrated and separated from the water by means of a mechanical rotary filter and subsequently expelled from the plant.

Based on a lab scale study, a pilot scale system for treatment of the clean-up water has been designed and implemented, with the aim to study a possible plant-scale process for the removal of dissolved tar, thus allowing wastewater recirculation in a closed loop with minimum make-up. Adsorption with carbon-based sorbents has been tested, with commercial activated carbon (AC) but also with two coal samples from South Africa and Venezuela (both characterised by similar properties: a lower heating value of 25 MJ/kg and a content of carbon, moisture and ash of 54%, 8% and 15% by weight, respectively), to assess the possibility to recirculate the exhaust sorbent in the gasification unit. The detailed results of this analysis at lab- and pilot-scale can be found elsewhere [43]. Lab scale experiments show that samples of coal have very low sorption capacity, one order of magnitude lower than the corresponding values of AC, and low retention times in the breakthrough curves.

Based on these results, the scaling up of the process to plant size has been assessed. A further experimental campaign has been carried out, where the pilot-scale column has been inserted in the secondary line and fed for several days, evaluating the turnover time of the sorbent with different flow rates. As a result, a turnover of about 50 h is required to ensure almost complete removal of dissolved tar with columns of 2.5–3.0 m height. The exhaust sorbent has been then characterised, in order to evaluate whether it can be mixed with the biomass in the feed to gasification. Table 4 shows the comparison between the characteristics of raw sorbent, exhausted sorbent, and a biomass already used in the feed of the gasification unit. The assessment shows that the exhaust AC can be added to the feed: heat power is higher than the corresponding amount of biomass, the content of ashes is similar. The moisture content is considerably higher, which can be a problem in a field application. However, AC may constitute a considerable fraction of the feed: 20% of AC results in a 2% increase of moisture in the feed, which can be acceptable for the plant. Moreover, a possible pre-treatment of drying may be implemented.

Table 4. Characterisation of raw and exhaust activated carbon (% by weight).

Parameter	Raw AC		Exhaust AC	
	As Received	Dry Basis	As Received	Dry Basis
Fixed carbon	79.07	82.03	56.08	79.80
Volatiles	2.89	3.00	6.60	9.40
Moisture	3.62	0.00	29.75	0.00
Ash	14.43	14.97	7.58	10.80
Lower heating value (MJ/kg, dry basis)	26.39	n.a.	18.47	n.a.

This study, in addition to all the previous experimental activities on the Sotacarbo demonstration-scale unit [39], has allowed confirmation of the efficiency of the fixed-bed updraft gasification process with a further improvement of the technology, with particular reference to the applications for power generation from waste biomass, such as wood chips from wood management but also agricultural residues. Most of the gasification results have been used to optimise a commercial-scale unit under design in Alaska [4].

Particular attention has been paid to develop and optimise the wastewater management; the introduction of the oil skimmer in the pilot unit reduced the organic load from 10% to 30%, depending of the runs. Moreover, several experiments have been carried out on the use of active carbon to treat the pyrolysis liquids (oils and tars), with the potential reuse (and energy recovery) of the spent material as primary fuel of the gasification unit. New tests are currently underway to assess the content of other contaminants that can be found in raw syngas generated by different feedstock and in different operating conditions. The stated goal is to make the necessary changes to connect the system to the electricity grid.

As an alternative solution, recovery of pyrolysis oils separated from raw syngas as a fuel additive in diesel engines will be exploited [4].

4. Conclusions

This paper reports the main results of the experimental activity—carried out on a demonstration plant at the Sotacarbo Research Centre in Southwest Sardinia (Italy)—to optimise and develop a fixed-bed updraft gasification process for power generation from biomass. In particular, the experimental campaign with stone pine and eucalyptus wood chips as primary fuels has shown a very high gasification performance, around 79–80% in terms of cold gas efficiency. Syngas has a very stable composition during the whole run (also thanks to the robust grate driving system, specifically designed and modified by ENEA and Sotacarbo to optimise the process operation), with a mean lower heating value in the order of 4.3–4.6 MJ/kg (syngas being composed of 52–53% by nitrogen from the gasification air).

Even if the maximum fuel bed temperature ranges between 820 and 880 °C, the highest zone of the fuel bed is relatively cold, with a freeboard mean temperature of about 300–380 °C. It involves a relatively high tar formation that required a careful design of the syngas cleaning section and particular attention to optimise the wastewater management to reduce make-ups and wastewater disposal and possibly recover the high energy content associated with tar and spent sorbents. A first characterisation of the process has been carried out and the modalities and key parameters to reach the standard operating conditions have been identified. Moreover, the optimal process parameters for the operation of the syngas cleaning section have been identified and the configuration has been modified (in particular, with the introduction of the oil skimmer integrated with the wet scrubbing system), with the result of a 60% reduction of wastewater disposal.

Author Contributions: Conceptualization, G.C., P.D., C.B. and M.M.; methodology, G.C., P.D., C.B. and M.M.; formal analysis, G.C., P.D., C.B., M.M. and S.M.; investigation, G.C., P.D., C.B. and S.M.; resources, P.D., M.M. and E.M.; data curation, G.C. and S.M.; writing—original draft preparation, G.C., P.D., C.B., S.M., E.M., M.M. and A.P.; writing—review and editing, C.B., P.D., G.C., M.M., E.M. and A.P.; supervision, P.D., M.M., E.M. and A.P.; project administration, P.D. and E.M.; funding acquisition P.D. and E.M. All authors have read and agreed to the published version of the manuscript.

Funding: This research has been funded by the Italian Ministry of Economic Development within the Research on Electric System programme (PAR2017, CUP: I12F17000070001).

Acknowledgments: The authors are grateful to Massimo Subrizi, Paolo Miraglia, Fabrizio Tedde and Davide Marotto for their support in the operation management of the demonstration plant and Alessando Orsini for his support in data acquisition and pre-processing.

Conflicts of Interest: The authors declare no conflict of interest. The funders had no role in the design of the study; in the collection, analyses, or interpretation of data; in the writing of the manuscript, or in the decision to publish the results.

References

1. International Energy Agency. *World Energy Outlook 2019*; International Energy Agency: Paris, France, 2019.

2. European Union. Directive (EU) 2018/2001 of the European Parliament and of the Council of 11 December 2018 on the promotion of the use of energy from renewable sources. *Official J. Eur. Union* **2018**, *61*, 82–209.

3. Krzyzaniak, M.; Stolarski, M.J.; Graban, L.; Lajszner, W.; Kuriata, T. Camelina and crambe oil crops for bioeconomy—Straw utilisation for energy. *Energies* **2020**, *13*, 1503. [CrossRef]

4. Ward, C.; Goldstein, H.; Maurer, R.; Thimsen, D.; Sheets, B.J.; Hobbs, R.; Isgrigg, F.; Steiger, R.; Revay Madden, D.; Porcu, A.; et al. Making coal relevant for small scale applications: Modular gasification for syngas/engine CHP applications in challenging environments. *Fuel* **2020**, *267*, 117303. [CrossRef]

5. Ferreira, S.; Minteiro, E.; Calado, L.; Silva, V.; Brito, P.; Vilarinho, C. Experimental and modeling analysis of brewers' spent grains gasification in a downdraft reactor. *Energies* **2019**, *12*, 4413. [CrossRef]

6. Marcantonio, V.; Bocci, E.; Monarca, D. Development of a chemical quasi-equilibrium model of biomass waste gasification in a fluidized-bed reactor by using Aspen Plus. *Energies* **2020**, *13*, 53. [CrossRef]

7. Gallucci, F.; Liberatore, R.; Sapegno, L.; Volponi, E.; Venturini, P.; Rispoli, F.; Paris, E.; Carnevale, M.; Colantoni, A. Influence of oxidant agent on syngas composition: Gasification of hazelnut shells through an updraft reactor. *Energies* **2020**, *13*, 102. [CrossRef]

8. Chambon, C.L.; Karia, T.; Sandwell, P.; Hallett, J.P. Techno-economic assessment of biomass gasification-based mini-grids for productive energy applications: The case of rural India. *Renew. Energy* **2020**, *154*, 432–444. [CrossRef]

9. Ren, J.; Liu, J.L.; Zhao, X.Y.; Cao, J.P. Methanation of syngas from biomass gasification: An overview. *Int. J. Hydrogen Energy* **2020**, *45*, 4223–4243. [CrossRef]

10. Szul, M.; Iluk, T.; Sobolewski, A. High-temperature, dry scrubbing of syngas with use of mineral sorbents and ceramic rigid filters. *Energies* **2020**, *13*, 1528. [CrossRef]

11. Situmorang, Y.A.; Zhao, Z.; Yoshida, A.; Abudula, A.; Guan, G. Small-scale biomass gasification systems for power generation (<200 kW class): A review. *Renew. Sustain. Energy Rev.* **2020**, *117*, 109486.

12. Siddiqui, H.; Thengane, S.K.; Sharma, S.; Mahajani, S.M. Revamping downdraft gasifier to minimize clinker formation for high-ash garden waste as feedstock. *Bioresour. Technol.* **2018**, *266*, 220–231. [CrossRef] [PubMed]

13. Sikarwar, V.S.; Zhao, M.; Clough, P.; Yao, J.; Zhong, X.; Memon, M.Z.; Shah, N.; Anthony, E.J.; Fennell, P.S. An overview of advances in biomass gasification. *Energy Environ. Sci.* **2016**, *9*, 2939. [CrossRef]

14. Porcu, A.; Sollai, S.; Marotto, D.; Mureddu, M.; Ferrara, F.; Pettinau, A. Techno-economic analysis of a small-scale biomass-to-energy BFB gasification-based system. *Energies* **2019**, *12*, 494. [CrossRef]

15. Kittivech, T.; Fukuda, S. Investigating agglomeration tendency of co-gasification between high alkali biomass and woody biomass in a bubbling fluidized bed system. *Energies* **2020**, *13*, 56. [CrossRef]

16. Huang, Y.; Zhao, Y.J.; Hao, Y.H.; Wei, G.Q.; Feng, J.; Li, W.Y.; Yi, Q.; Mohamed, U. A feasibility analysis of distributed power plants from agricultural residues resources gasification in rural China. *Biomass Bioenergy* **2019**, *121*, 1–12. [CrossRef]

17. Ren, J.; Cao, J.P.; Zhao, X.Y.; Yang, F.L.; Wei, X.Y. Recent advances in syngas production from biomass catalytic gasification: A critical review on reactors, catalysts, catalytic mechanisms and mathematical models. *Renew. Sustain. Energy Rev.* **2019**, *116*, 109426. [CrossRef]

18. Lee, S.Y.; Park, S.W.; Alam, M.T.; Jeong, J.O.; Seo, Y.C.; Choi, H.S. Studies on the gasification performance of sludge cake pre-treated by hydrothermal carbonization. *Energies* **2020**, *13*, 1442. [CrossRef]

19. Singh Siwal, S.; Zhang, Q.; Sun, C.; Thakur, S.; Gupta, V.K.; Thakur, V.K. Energy production from steam gasification processes and parameters that contemplate in biomass gasifier—A review. *Bioresour. Technol.* **2020**, *297*, 122481. [CrossRef]

20. Hobbs, M.L.; Radulovic, P.T.; Smoot, L.D. Modeling fixed-bed coal gasifier. *AIChE J.* **1992**, *38*, 681–702. [CrossRef]

21. Hahn, O.H.; Wesley, P.D.; Swisshelm, B.A.; Maples, S.; Withrow, J. A mass and energy balance of a Wellman–Galusha gasifier. *Fuel Process. Technol.* **1979**, *2*, 332–334. [CrossRef]

22. Yu, J.; Smith, J.D. Validation and application of a kinetic model for biomass gasification simulation and optimization in updraft gasifiers. *Chem. Eng. Process. Process Intensif.* **2018**, *125*, 214–226. [CrossRef]

23. Mehta, V.; Chavan, A. Physico-chemical treatment of tar-containing wastewater generated from biomass gasification plants. *Int. J. Chem. Mol. Eng.* **2009**, *3*, 458–465.

24. Manek, B.; Javia, M.S.; Harichandan, A.; Ramani, H. A CFD based comprehensive study of coal-fired updraft gasifier in ceramic industry. *Therm. Sci. Eng. Prog.* **2019**, *9*, 11–20. [CrossRef]

25. Murgia, S.; Vascellari, M.; Cau, G. Comprehensive CFD model of an air-blown coal-fired updraft gasifier. *Fuel* **2012**, *101*, 129–138. [CrossRef]

26. Lu, D.; Yoshikawa, K.; Ismail, T.M.; El-Salam, M.A. Assessment of the carbonized woody briquette gasification in an updraft fixed bed gasifier using the Euler-Euler model. *Appl. Energy* **2018**, *220*, 70–86. [CrossRef]

27. Lu, D.; Yoshikawa, K.; Ismail, T.M.; El-Salam, M.A. Computational fluid dynamics model on updraft gasifier using carbonized woody briquette as fuel. *Energy Procedia* **2017**, *142*, 166–171. [CrossRef]

28. James R, A.M.; Yuan, W.; Wang, D.; Wang, D.; Kumar, A. The Effect of Gasification Conditions on the Surface Properties of Biochar Produced in a Top-Lit Updraft Gasifier. *Appl. Sci.* **2020**, *10*, 688. [CrossRef]

29. Lei, M.; Hai, J.; Cheng, J.; Lu, J.; Zhang, J.; You, T. Variation of toxic pollutants emission during a feeding cycle from an updraft fixed bed gasifier for disposing rural solid waste. *Chin. J. Chem. Eng.* **2018**, *26*, 608–613. [CrossRef]

30. Font Palma, C. Modelling of tar formation and evolution for biomass gasification: A review. *Appl. Energy* **2013**, *111*, 129–141. [CrossRef]

31. Devi, J.; Ptasinski, K.J.; Janssen, F.J.J.G. A review of the primary measures for tar elimination in biomass gasification processes. *Biomass Bioenergy* **2003**, *24*, 125–140. [CrossRef]

32. Mahjoub, B.; Jayr, E.; Bayard, R.; Gourdon, R. Phase partition of organic pollutants between coal tar and water under variable experimental conditions. *Water Rese.* **2000**, *34*, 3551–3560. [CrossRef]

33. Ruiz, J.; Juárez, M.; Morales, M.; Muñoz, P.; Mendívil, M. Biomass gasification for electricity generation: Review of current technology barriers. *Renew. Sustain. Energy Rev.* **2013**, *18*, 174–183. [CrossRef]

34. Park, H.I.; Wu, C.J.; Lin, L.S. Coal tar wastewater treatment and electricity production using a membrane-less tubular microbial fuel cell. *Biotechnol. Bioprocess Eng.* **2012**, *17*, 654–660. [CrossRef]

35. Luthy, R.G.; Dzombak, D.A.; Peters, C.A.; Roy, S.B.; Ramaswami, A.; Nakles, D.V.; Nott, B.R. Remediating tar-contaminated soils at manufactured gas plant sites-technological challenges. *Environ. Sci. Technol.* **1994**, *28*, 266–276. [CrossRef]

36. Milne, T.A.; Evans, R.J.; Abatzoglou, N. *Biomass Gasifier "Tars": Their Nature, Formation, and Conversion*; Report NREL/TP-570-25357; U.S. National Renewable Energy Laboratory: Golden, CO, USA, 1998. Available online: https://www.nrel.gov/docs/fy99osti/25357.pdf (accessed on 1 October 2019).

37. Han, J.; Kim, H. The reduction and control technology of tar during biomass gasification/pyrolysis: An overview. *Renew. Sustain. Energy Rev.* **2008**, *12*, 397–416. [CrossRef]

38. Nyazi, K.; Baçaoui, A.; Yaacoubi, A.; Darmstadt, H.; Adnot, A.; Roy, C. Influence of carbon black surface chemistry on the adsorption of model herbicides from aqueous solution. *Carbon* **2005**, *43*, 2218–2221. [CrossRef]

39. Calí, G.; Deiana, P.; Bassano, C.; Maggio, E. Experimental activities on Sotacarbo 5 MWth gasification demonstration plant. *Fuel* **2017**, *207*, 671–679. [CrossRef]
40. Ghani, M.U.; Radulovic, P.T.; Smoot, L.D. An improved model for fixed-bed coal combustion and gasification: Sensitivity analysis and applications. *Fuel* **1996**, *75*, 1213–1226. [CrossRef]
41. Thimsen, D.P.; Maurer, R.E.; Pooler, A.R.; Pui, D.Y.H.; Liu, B.Y.H.; Kittelson, D.B. Fixed-Bed Gasification Research Using US Coals. Volume 19—Executive Summary. US Bureau of Mines Contract H0222001. 1985. Available online: https://www.osti.gov/servlets/purl/5704367 (accessed on 15 May 2019). [CrossRef]
42. De Filippis, P.; Scarsella, M.; de Caprariis, B.; Uccellari, R. Biomass gasification plant and syngas clean-up system. *Energy Procedia* **2015**, *75*, 240–245. [CrossRef]
43. Calí, G.; Deiana, P.; Maggio, E.; Marotto, D.; Mascia, M.; Vacca, A. Management and treatment of the clean-up water from the scrubber of a coal and biomass gasification plant: An industrial case study. *Chem. Eng. Trans.* **2019**, *74*, 337–342.
44. Conte, L.; Falletti, L.; Zaggia, A.; Milan, M. Polyfluorinated organic micropollutants removal from water by ion exchange and adsorption. *Chem. Eng. Trans.* **2015**, *43*, 2257–2262.
45. Liu, Z.Q.; You, L.; Xiong, X.; Wang, Q.; Yan, Y.; Tu, J.; Cui, Y.H.; Li, X.Y.; Wen, G.; Wu, X. Potential of the integration of coagulation and ozonation as a pretreatment of reverse osmosis concentrate from coal gasification wastewater reclamation. *Chemosphere* **2019**, *222*, 696–704. [CrossRef] [PubMed]

Article

Experimental and Modeling Analysis of Brewers´ Spent Grains Gasification in a Downdraft Reactor

Sérgio Ferreira [1], Eliseu Monteiro [2,3,*], Luís Calado [2], Valter Silva [2], Paulo Brito [2] and Cândida Vilarinho [1]

[1] CT2M—Centre for Mechanical and Materials Technologies, Mechanical Engineering Department of Minho University, 4804-533 Guimarães, Portugal; sergio.c.m.ferreira@gmail.com (S.F.); candida@dem.uminho.pt (C.V.)
[2] VALORIZA—Research Center for Endogenous Resource Valorisation, Polytechnic Institute of Portalegre, 7300-555 Portalegre, Portugal; luis.calado@ipportalegre.pt (L.C.); valter.silva@ipportalegre.pt (V.S.); pbrito@ipportalegre.pt (P.B.)
[3] CIENER-LAETA/Faculty of Engineering, University of Porto, 4200-465 Porto, Portugal
[*] Correspondence: elmmonteiro@portugalmail.pt or eliseu@ipportalegre.pt; Tel.: +351-254-300-200

Received: 12 October 2019; Accepted: 16 November 2019; Published: 20 November 2019

Abstract: The first part of the current reported work presents experimental results of brewers' spent grains gasification in a pilot-scale downdraft gasifier. The gasification procedure is assessed through various process characteristics such as gas yield, lower heating value, carbon conversion efficiency, and cold gas efficiency. Power production was varied from 3.0 to 5.0 kWh during the gasification experiments. The produced gas was supplied to an internal combustion engine coupled to a synchronous generator to produce electricity. Here, 1.0 kWh of electricity was obtained for about 1.3 kg of brewers' spent grains pellets gasified, with an average electrical efficiency of 16.5%. The second part of the current reported work is dedicated to the development of a modified thermodynamic equilibrium model of the downdraft gasification to assess the potential applications of the main Portuguese biomasses through produced gas quality indices. The Portuguese biomasses selected are the main representative forest residues (pine, eucalyptus, and cork) and agricultural residues (vine prunings and olive bagasse). A conclusion can be drawn that, using air as a gasifying agent, the biomass gasification provides a produced gas with enough quality to be used for energy production in boilers or turbines.

Keywords: autothermal gasification; downdraft reactor; thermodynamics; chemical equilibrium; carbon boundary point

---

1. Introduction

Population and incomes rising will continue to push up the global energy demand, according to the International Energy Agency (IEA) [1]. Additionally, the limited availability of energy resources calls for new advantageous and creative solutions for safe energy supply. Bioenergy is considered one of the key options of the renewable energy field to mitigate greenhouse gas emissions, replace fossil fuels, and ensuring a more secure and sustainable energy system [2].

Biomass conversion via gasification is an established technique in modern bioenergy systems [3]. It is an important process to convert biomass into a combustible gas that can be used in boilers, turbines, engines, and even fuel cells. This combustible gas can also be used as a raw material in the production of synthetic fuels or chemicals [4,5].

Lignocellulosic biomass represents the most available renewable resource on the planet [6]. The interest in using lignocellulosic biomass as a renewable resource for bioproducts production is rising, especially due to their abundance, low cost, and their production does not compete with the food chain [7]. Among the lignocellulosic biomasses, spent grain [8] has received increased interest in the last few years [9–14].

Spent gain is the major byproduct of the brewing process that includes spent yeast and spent hops [14]. Basically, beer is a yeast fermenting product of the brewer wort obtained from malted barley, sometimes combined with other cereals called adjuncts (maize, rice, oats, wheat, etc.), with the addition of hops [15].

Spent grain is generated in the beer-brewing process, which begins with the production of the wort. The wort comprises crushed barley malt mixed with water in a mash tun. The temperature is gradually increased to about 78 °C in order to transform the malt starch into fermentable and non-fermentable sugars. The insoluble undegraded part of the barley malt grain obtained in the mixture with the wort at the end of this process is known as spent grain. Spent grain is the most abundant brewing byproduct, corresponding to around 85% of total byproducts generated [9].

During the brewing process, the wort should be submitted to the boiling stage, with the purpose of hop addition and the extraction of its aroma and bitterness compounds [16]. In this step, the wort loses part of its high nitrogen content due to the formation of a precipitate called hot trub or spent hop. Spent hop is the second solid residue generated in the brewing process, which results principally from insoluble coagulation of high molecular weight proteins. Comprising around 2% of the total byproducts generated during brewing, spent hop is the lowest byproduct of the brewing process. The main use of spent yeast is as animal feed as a source of protein and water-soluble vitamins [17].

The fermentation stage is triggered by the addition of yeast to the filtered wort, converting sugar to alcohol and carbon dioxide. Before full maturation of the beer, the excess yeast is collected and can be re-used in the brewing process as many as six times. After this, it becomes brewer's spent yeast. Comprising around 13% of the total byproducts generated during brewing, spent yeast is the second biggest byproduct of the brewing process. Spent yeast is an interesting byproduct since it contains a high level of nutrients, and several technologies exist that can transform this byproduct into a valuable resource. However, to date, its industrial utilization is very limited because of the fast contamination and spoilage of spent yeast as a result of the activity of microorganisms. The bitterness of spent hops does not make it a good candidate for use as an animal feed [17]. The main methods of disposal are to reuse them as fertilizer or compost [18].

From this brief overview of the brewing process, it is possible to verify that the brewer's spent grains (BSG) are the most representative byproduct of the brewing process. The brewing sector in Portugal generates around 135,000 tonnes of BSG per year [13], which are mainly used as animal feed. Nevertheless, recent developments exposed other possible applications such as the production of various value-added bio-products [9–11] and energy generation for the brewing process [11–13].

There are numerous methods of exploiting biomass to produce energy and fuels. Among them, gasification processes seem to be a good option for small- to large-scale applications since the sub-stoichiometric conditions in the reactor allows for much lower pollutant emissions than combustion processes [19,20].

Literature is very scarce on the subject of the thermochemical conversion of BSG through pyrolysis and gasification processes. Mahmood et al. [21] used a batch pyrolysis reactor to pyrolyze small samples of BSG. A reforming nickel catalyst was added downstream of the reactor for cracking and reforming of the pyrolysis products with and without the addition of steam. The obtained results indicated that catalytic reforming promotes an increase in CO and H_2 contents. The process also showed an increase in heating value for the produced gas as the reforming temperature increased. Borel et al. [22] performed thermogravimetric studies on the pyrolysis of BSG to evaluate its potential for bio-oil production. The results suggest a good potential of BSG for bio-oil production due to their high heating value and high volatile matter content. Ulbricha et al. [11] studied the influence of temperature and residence time on the hydrothermal carbonization and carbon dioxide gasification of brewers' spent grains. The results suggest that prolonged residence times and higher temperature decreases energy and mass yields and increases heating values and fixed carbon formation in the coal. Carbon dioxide reaction rates of chars after pyrolysis are decreased due to the formation of fixed carbon during the hydrothermal carbonization. Activation energies of the carbon dioxide reaction also decrease with higher hydrothermal carbonization reaction rates. Ferreira et al. [13] performed a very complete characterization of brewers' spent grains and subject it to steam gasification in an allothermal batch reactor. BSG was characterized through proximate, ultimate, and thermogravimetric analysis and a van Krevelen diagram. The results suggest that BSG has similar characteristics to common lignocellulosic biomasses. BSG steam gasification was carried out to determine the influence of temperature and the steam-to-biomass ratio on the produced gas composition. They found that CO and CH_4 contents decrease with the steam-to-biomass ratio, while H_2 and CO_2 contents increase. The temperature increase leads to increased CO and H_2 contents and decreased CH_4 and CO_2 contents.

Another possible route of valorization for BSG is downdraft gasification, which is a proven technology and a low-cost process with the additional advantage of generating very low tar levels [19]. The produced gas can be subjected to cogeneration, for which BSG gasification behavior is still unknown. Therefore, the first part of this paper is dedicated to the study of the influence of some process conditions on the BSG downdraft gasification using a power pallet downdraft gasifier from All Power Labs (Berkeley, CA, USA).

The second part of this paper is dedicated to the development and implementation of a mathematical model to understand and predict the BSG downdraft gasification process performance and to assess the influence of diverse variables on the process performance for other biomass substrates. The main reason for that is to take advantage of the possibility provided by numerical models in order to circumvent time-consuming and costly experimental trials [23,24].

Gasification modeling and simulation may be achieved through different approaches, such as equilibrium models, kinetic models, computational fluid dynamics, and artificial neural networks [24,25]. Equilibrium models have the capacity to predict the maximum possible yield of a product; hence, they are not so accurate. However, thermodynamic models may be more suitable for some applications, given that they are independent of the gasifier's design and do not include any information about conversion mechanisms [26]. Therefore, they are the best choice for preliminary studies and parametric studies [25,27].

There are many modeling studies on equilibrium modeling of lignocellulosic biomass such as wood [28,29], agriculture residues [29,30], or pine [31]. However, to the best of our knowledge, there are no modeling studies on BSG downdraft gasification. Therefore, the second part of this work is dedicated to the development of a modified thermodynamic equilibrium model of the downdraft gasification to assess the potential application of the main Portuguese biomasses through produced gas quality indices.

2. Materials and Methods

2.1. Brewer's Spent Grain Characterization

The BSG used in downdraft gasification experiments was characterized in a previous study [13] in terms of ultimate and proximate analysis and heating value. The main results of BSG characterization are shown in Table 1.

Table 1. Characterization of brewers' spent grain pellets [13].

Parameter	Value	Parameter	Value
Moisture (%)	12.7	Density (kg/m^3)	517
HHV (MJ/kg)	17.8	LHV (MJ/kg)	16.5
Proximate analysis (%, db)		Ultimate analysis (%, daf)	
Ash	3.8	C	48.3
Volatile	86.8	H	5.6
Fixed carbon	9.4	N	5.5

2.2. Pilot-Scale Downdraft Reactor

A pilot-scale integrated gasification power system—a 20 kW Power Pallet (PP20) supplied by All Power Labs, Berkeley, CA, USA—was used. This system is mainly comprised of a downdraft reactor, an internal combustion engine, an electrical synchronous generator, and a process control unit. The external appearance of the gasifier and the main specifications are given in Figure 1 and Table 2, respectively.

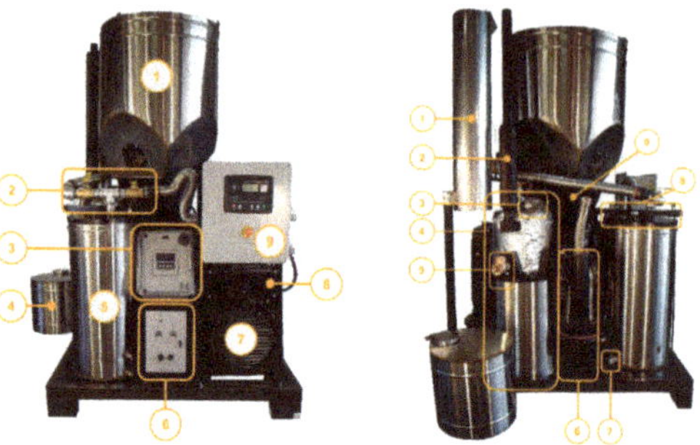

Figure 1. Pilot-scale downdraft reactor. Left: (1) Hopper, (2) Valves to flare and engine, (3) Control unit, (4) Ash vessel, (5) Gas filter, (6) Operational panel, (7) Generator, (8) Wiring box, (9) Grid-tie. Right: (1) Flare, (2) Exhaust stack, (3) Reactor access port, (4) Gasifier, (5) Air inlet check valve, (6) Cyclone, (7) Filter condensate drain bung, (8) Filter lid-locking lever, (9) drying bucket.

Table 2. Power Pallet specifications [32].

Property	Value
Power output	3–15 kW at 50 Hz
Biomass consumption	18 kg/h at 15 kW
Moisture tolerance	<30%
Dimensions	$1.4 \times 1.4 \times 2.2$ m
Weight	1065 kg
Feedstock hopper capacity	0.33 m^3

The downdraft gasifier is made of stainless steel, and its kernel is made of coated ceramic. It is comprised of four sections corresponding to the different gasification phases (drying, pyrolysis, combustion and cracking, and reduction), as depicted in Figure 2. The power pallet system operates at a negative pressure to avoid gas leaks. Therefore, the hopper is sealed to maintain the negative pressure. In the drying zone, a heat exchanger with the hot departing gas reduces the moisture content of the biomass. A worm screw carries the dried biomass for the downdraft reactor. Drying and pyrolysis are both endothermic phases of gasification. Therefore, the power pallet includes a physical separation between these gasification phases to avoid competition for the heat required to each phase and reduce the amount of water in the gasifier, which would tend to agglomerate the tars and the soot into droplets and hamper their elimination by thermal cracking.

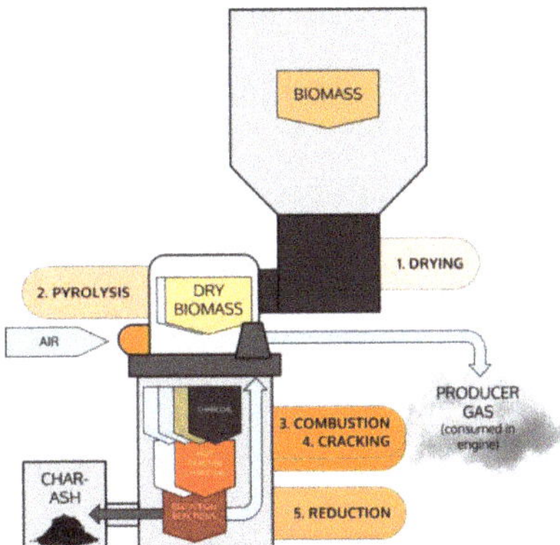

Figure 2. Schematic view of the gasifier system.

The heat required for the pyrolysis that occurs at the top of the gasifier comes from combustion reactions that take place in the middle of the gasifier and from a heat exchanger with the engine's exhaust gases. The air intake flow to the combustion zone also experiences a heat exchange with the departing gas. In the reduction zone at the bottom of the gasifier, a grate allows ash and char granules to pass. This grate is shaken by the system to smooth the passage of small granules and ash to the bottom of the gasifier, which, in turn, facilitates the flow of the produced gas.

The produced gas leaves the gasifier and enters into a cyclone that precipitates larger particles and condensates present in the produced gas stream. After that, the produced gas can follow two different routes, depending on the operating conditions. During start-up, the produced gas follows the flare route, as is the low temperature in the gasifier does not permit it to crack the produced tars; that would damage the internal combustion engine. When the temperature in the reactor stabilizes, the produced gas follows the internal combustion engine route. Meanwhile, the produced gas goes through a packed bed filter, which ultimately removes moisture and other contaminants.

The internal combustion engine is a spark-ignition engine (GM Vortec type) (General Motors, Detroit, MI, USA) properly modified to use low calorific combustible gases. The air-fuel ratio is tuned through a control unit and an oxygen (lambda) sensor. The equivalence ratio is seen on the control unit display, which allows the user to verify that the air-fuel mixture is generally stoichiometric [33].

The Power Pallet has a direct connection between the engine's drive shaft and the generator. For the generator to output electricity with a constant frequency of 50 Hz, it also has an engine governor to ensure that the engine turns at 1500 rpm (synchronous generator with four poles) while varying the power output to match the load on the generator. Further details about the pilot-scale integrated gasification power production system and its equipment can be found elsewhere [32,34,35].

2.3. Experimental Procedure

Prior to the gasification experiments, the reactor was fully emptied and then filled with BSG pellets. For the BSG gasification trials, the hopper was filled with 50 kg of BSG pellets working in a batch mode. The system was initially tested to avoid gas leaking from the gasifier during operation. The start-up of the gasifier was done by a propane burner. When the temperature rises above 700 °C, the produced gas is supplied to the internal combustion engine instead of the flare. At the end of each gasification trial, the reactor, gas filter, grate basket, and ash container were cleaned.

The produced gas was sampled at the exit of the gas filter using Tedlar bags (CEL Scientific Corp., Cerritos, CA, USA) in intervals of 15 min during the one-hour test for each operational condition. Therefore, four produced gas samples were taken for each operational condition and analyzed in a Varian 450-GC (Scion, Austin, TX, USA) gas chromatograph with two thermal conductivity detectors, enabling the recognition of CO, CO_2, CH_4, C_2H_2, C_2H_4, C_2H_6, H_2, O_2, and N_2 using nitrogen and helium as carrier gases. The experimental results presented in this paper are the average of these four samples.

2.4. Mathematical Model

The modified stoichiometric equilibrium model presented herein is based on the carbon boundary point (CBP) concept. The CBP is attained when enough gasifying agent is supplied to achieve complete gasification [36,37]. Therefore, it is generally considered to define the optimal conditions of a gasification process [36,38]. It is also considered a two-stage model. In the first stage, below the carbon boundary point, only heterogeneous reactions take place. In the second stage, above the carbon boundary point, only homogeneous reactions occur [31,36].

The modified stoichiometric equilibrium model presented herein is based on the following typical assumptions [27,36]:

- The gasifier is considered zero-dimensional and adiabatic;
- Residence time is long enough for the equilibrium state to be achieved;
- Hydrodynamic behavior is considered as homogeneous mixing with uniform pressure and temperature;
- Tars and ashes contents are considered negligible.

The stoichiometric equilibrium model is developed as a two-stage model considering a sub-model for gasification at and below the CBP, where a heterogeneous equilibrium is present, and another sub-model for gasification above the CBP, where homogeneous equilibrium is present, i.e., all the components are in the gaseous state, as in references [31,38].

2.4.1. Model at and below the CBP

- Mass Balance

The global gasification reaction of a mole of biomass in m moles of air can be expressed as follows:

$$CH_xO_yN_z + wH_2O + m(O_2 + 3.76N_2) \rightarrow n_{H_2}H_2 + n_{CO}CO + n_{H_2O}H_2O + n_{CO_2}CO_2 + n_{CH_4}CH_4 + n_{N_2}N_2 + n_{char}Char \tag{1}$$

where the subscripts n_i denotes the stoichiometric coefficients. $CH_xO_yN_z$ denotes the biomass material, and x, y, and z denote the numbers of atoms of hydrogen, oxygen, and nitrogen per number of atoms of carbon present in the biomass. w and m denote the quantity of moisture and oxygen per mole of biomass, respectively. The subscripts x, y, z, and w are obtained from the ultimate analysis of the biomass.

The atomic balance for the chemical elements C, H, O, and N are defined as follows:

$$C: n_{CO} + n_{CO_2} + n_{CH_4} + n_{char} = 1 \tag{2}$$

$$H: 2n_{H_2} + 2n_{H_2O} + 4n_{CH_4} = x + 2w \tag{3}$$

$$O: n_{CO} + n_{H_2O} + 2n_{CO_2} = y + w + 2m \tag{4}$$

$$N: 2n_{N_2} = z + 7.52m \tag{5}$$

- Thermodynamic Heterogeneous Equilibrium

Three independent equilibrium chemical reactions are enough for the heterogeneous equilibrium. The relevant gasification reactions in this regard are the Boudouard reaction (Equation (6)), the water–gas (Equation (7)), and methane formation (Equation (8)) [31,39].

$$C + CO_2 \leftrightarrow 2CO \tag{6}$$

$$C + H_2O \leftrightarrow CO + H_2 \tag{7}$$

$$C + 2H_2 \leftrightarrow CH_4 \tag{8}$$

The equilibrium constants for those reactions are [31]

$$K_6(T) = \frac{\left(\frac{n_{CO}}{n_t}\right)^2}{\left(\frac{n_{CO_2}}{n_t}\right)}\left(\frac{P_{ref}}{P}\right) \tag{9}$$

$$K_7(T) = \frac{\left(\frac{n_{CO}}{n_t}\right)\left(\frac{n_{H_2}}{n_t}\right)}{\left(\frac{n_{H_2O}}{n_t}\right)}\left(\frac{P_{ref}}{P}\right) \tag{10}$$

$$K_8(T) = \frac{\frac{n_{CH_4}}{n_t}}{\left(\frac{n_{H_2}}{n_t}\right)^2}\left(\frac{P_{ref}}{P}\right) \tag{11}$$

where n_t denotes the total number of moles of produced gas, P_{Ref} denotes the standard pressure (1 atm), P denotes the pressure at the operating condition, and $k_i(T)$ denotes the equilibrium constant that can also be obtained using the standard Gibbs function [39],

$$\ln K_i = -\frac{\sum_{i=1}^{N} n_i \Delta g_{f,T,i}^0}{RT} \tag{12}$$

where R denotes the universal gas constant and $\Delta g_{f,T,i}^0$ the standard Gibbs function of formation of the gas species i, which can be determined as follows [39]:

$$\Delta g_{f,T,i}^0 = h_f^0 - a'T\ln(T) - b'T^2 - \left(\frac{c'}{2}\right)T^3 - \left(\frac{d'}{3}\right)T^4 - \left(\frac{e'}{2T}\right) - f' - g'T \tag{13}$$

The coefficients a'–g' and the enthalpy of formation of the gases are given in Table 3.

Table 3. Enthalpy of formation (kJ/mol) and coefficients of the Gibbs equation (kJ/mol) [40].

Substance	h^0_f	a'	b'	c'	d'	e'	f'	g'
CO	−110.5	5.619×10^{-3}	-1.190×10^{-5}	6.383×10^{-9}	-1.846×10^{-12}	-4.891×10^2	8.684×10^{-1}	-6.131×10^{-2}
CO$_2$	−393.5	-1.949×10^{-2}	3.122×10^{-5}	-2.448×10^{-8}	6.946×10^{-12}	-4.891×10^2	5.270	-1.207×10^{-1}
CH$_4$	−74.8	-4.620×10^{-2}	1.130×10^{-5}	1.319×10^{-8}	-6.647×10^{-12}	-4.891×10^2	1.411×10^1	-2.234×10^{-1}
H$_2$O	−241.8	-8.950×10^{-3}	-3.672×10^{-6}	5.209×10^{-9}	-1.478×10^{-12}	0.000	2.868	-1.722×10^{-2}

- Energy Balance

The gasification temperature can be obtained by the following global energy balance equation for 1 kg of biomass considering that the process is adiabatic [39].

$$\sum_i n_i \left[h_{f,i}^0 + \Delta H_{298}^T\right]_{i,\ reactants} = \sum_i n_i \left[h_{f,i}^0 + \Delta H_{298}^T\right]_{i,\ products} \tag{14}$$

Taking into account the global gasification reaction of Equation (1), the global energy balance can be expressed as

$$
\begin{aligned}
h^0_{f,biomass} &+ w\left(h^0_{f,H_2O} + h_{vap}\right) + m\, h^0_{f,O_2} + 3{,}76\, m\, h^0_{f,N_2} \\
&= n_{H_2} h^0_{f,H_2} + n_{CO} h^0_{f,CO} + n_{H_2O} h^0_{f,H_2O} + n_{CO_2} h^0_{f,CO_2} + n_{CH_4} h^0_{f,CH_4} \\
&+ n_{N_2} h^0_{f,N_2} + n_{char} h^0_{f,char} \\
&+ \left(n_{H_2} c_{p,H_2} + n_{CO} c_{p,CO} + n_{H_2O} c_{p,H_2O} + n_{CO_2} c_{p,CO_2} + n_{CH_4} c_{p,CH_4}\right. \\
&\left. + n_{N_2} c_{p,N_2} + n_{char} c_{p,char}\right)\Delta T
\end{aligned}
\tag{15}
$$

$h^0_{f,i}$, h_{vap}, and $c_{p,i}$ denote the biomass enthalpy of formation, the enthalpy of vaporization of water, and the specific heat, respectively. ΔT refers to the temperature difference at any given T and at 298 K. The enthalpy of formation of the biomass can be computed by the following relationship [41]:

$$
h^0_{f,biomass} = LHV + \sum_i \left[n_i h^0_{f,i}\right]_{i,products}
\tag{16}
$$

where $h^0_{f,biomass}$ denotes the enthalpy of formation of product i under complete combustion of the biomass and LHV denotes the lower heating value of the biomass. The LHV of the biomass is computed by subtracting the higher heating value (HHV) of the biomass by the enthalpy of vaporization of water as follows [26]:

$$
LHV_{bio} = HHV_{bio} - 2260 \times (0.09\,H + 0.01M)\left(\frac{kJ}{kg}\right)
\tag{17}
$$

where H and M denote the hydrogen and moisture fractions on an as-received basis. The value 2260 is the latent heat of the water in kJ/kg. The HHV of the biomass is computed accordingly to the correlation of Channiwala and Parikh [42],

$$
HHV_{bio} = 349.1\,C + 1178.3\,H + 100.5\,S - 103.4\,O - 15.1\,N - 21.1\,Ash\left(\frac{kJ}{kg}\right)
\tag{18}
$$

where the mass percentages of the compounds are those obtained by ultimate analysis on a dry basis.

Cp denotes the specific heat at constant pressure in kJ/kmol K that can be computed by the following empirical equation [41]:

$$
C_p(T) = a_1 + a_2 T + a_3 T^2 + a_4 T^{-2}
\tag{19}
$$

where the coefficients a_i are given in Table 4 for the chemical species involved.

Table 4. Coefficients for the specific heat calculation [41].

Species	a_1	a_2	a_3	a_4
C	16.336	0.60972×10^{-2}	-0.64762×10^{-6}	$-836{,}340$
CO	28.448	0.23633×10^{-2}	-0.24877×10^{-6}	4291.9
CO_2	36.299	0.20352×10^{-1}	-0.21455×10^{-5}	$-449{,}100$
CH_4	23.607	0.49622×10^{-1}	-0.52248×10^{-5}	$-212{,}800$
H_2O	28.166	0.14667×10^{-1}	-0.15433×10^{-5}	$100{,}230$
H_2	25.310	0.82575×10^{-2}	-0.86850×10^{-6}	$106{,}010$
N_2	27.883	0.29838×10^{-2}	-0.31384×10^{-6}	$38{,}452$

2.4.2. Model Above the CBP

- Mass Balance

The balance for, C, H_2, O_2, and N_2 can be given by [31]

$$C: \quad n_{gas}\left(n_{CO} + n_{CO_2} + n_{CH_4}\right)_{CBP} = n_{gas}\left(n_{CO} + n_{CO_2} + n_{CH_4}\right) \tag{20}$$

$$H_2: \quad n_{gas}\left(n_{H_2} + n_{H_2O} + 2n_{CH_4}\right)_{CBP} = n_{gas}\left(n_{H_2} + n_{H_2O} + 2n_{CH_4}\right) \tag{21}$$

$$O_2: n_{gas}\left[0.5\left(n_{CO} + n_{H_2O}\right) + n_{CO_2}\right]_{CBP} + n_{air} \times n_{O_{2,air}} + 0.5 \times n_{H_2O} = n_{gas}\left[0.5\left(n_{CO} + n_{H_2O}\right) + n_{CO_2}\right] \tag{22}$$

$$N_2: \quad n_{gas} \times n_{N_2, \; CBP} + n_{air} \times n_{N_2,air} = n_{gas} \times n_{N_2} \tag{23}$$

where n_{gas}, n_{CO}, n_{CO_2}, n_{CH_4}, n_{H_2} and n_{H_2O} denotes the molar amount of produced gas, carbon monoxide, carbon dioxide, methane, hydrogen, and water, respectively. n_{air}, $n_{O_2, \; air}$, and $n_{N_2, \; air}$ denotes the molar amount of air, oxygen in the air, and nitrogen in the air. The subscript *CBP* stands for a species molar amount at the CBP.

- Thermodynamic Homogeneous Equilibrium

The pertinent chemical reactions are the water–gas shift reaction (Equation (24)) and methanation reaction (Equation (25)) [36].

$$CO + H_2O \leftrightarrow CO_2 + H_2 \tag{24}$$

$$CO + 3H_2 \leftrightarrow CH_4 + H_2O \tag{25}$$

The equilibrium constants for those reactions are [31]

$$K_{24}(T) = \frac{\left(\frac{n_{H_2}}{n_t}\right) \times \left(\frac{n_{CO_2}}{n_t}\right)}{\left(\frac{n_{CO}}{n_t}\right) \times \left(\frac{n_{H_2O}}{n_t}\right)} \tag{26}$$

$$K_{25}(T) = \frac{\left(\frac{n_{H_2O}}{n_t}\right) \times \left(\frac{n_{CH_4}}{n_t}\right)}{\left(\frac{n_{CO}}{n_t}\right) \times \left(\frac{n_{H_2}}{n_t}\right)^3}\left(\frac{P_{ref}}{P}\right)^2 \tag{27}$$

These equilibrium constants are obtained using Equation (12).

- Energy Balance

The global energy balance equation for the homogenous stage of the model is defined as follows [31,38]:

$$\left(\Delta h_{gas} + LHV_{gas}\right)_{CBP} \times n_{gas,CBP} + \Delta h_{air} \times n_{air} + \Delta h_{water} \times n_{water} = \left(LHV_{gas} + \Delta h_{gas}\right) \times n_{gas} \tag{28}$$

where $\Delta h_{air} \times n_{air}$ denotes the product between the air enthalpy difference by the molar amount of air, $\Delta h_{water} \times n_{water}$ denotes the product between the water enthalpy difference by the molar amount of water vapor, Δh_{gas} denotes the produced gas enthalpy difference, and LHV_{gas} denotes the lower heating value of the produced gas. The unknowns in Equation (28) are computed thanks to Equations (16)–(19).

2.4.3. Calculation Procedure

It is known that equilibrium models at relatively low gasification temperatures overestimate carbon monoxide and hydrogen yields and underestimates carbon dioxide and methane yields [25,43]. Therefore, to improve the predictive capabilities of the developed equilibrium model, the multiplicative factors of Jarungthammachote and Dutta [39] were used. According to their methodology, the equilibrium constants of the water-gas reaction (Equation (10)) and methanation reaction (Equation (11)) were multiplied by 0.91 and 11.28, respectively. The values of n_{CO}, n_{CO_2}, n_{CH_4}, n_{H_2}, n_{H_2O}, n_t, and T_{CBP} are computed in the first stage of the model, i.e., below and at the CBP, assuming an initial temperature. In the second stage of the model, i.e., above the CBP, the same methodology was followed to solve the homogeneous equilibrium. The values of n_{CO}, n_{CO_2}, n_{CH_4}, n_{H_2}, and n_{H_2O}. obtained in the first stage of the model are used as input parameters for the second stage of the model.

The described stoichiometric modified equilibrium model was implemented and solved in Matlab (MathWorks Inc., Natick, MA, USA) software using the Newton–Raphson method.

3. Results and Discussion

3.1. Experimental Results

The operating conditions of the downdraft gasifier were characterized by the biomass feed rate, air feed rate, equivalence ratio (ER), and produced gas composition. Table 5 shows the operating conditions, averaged produced gas fractions, and efficiencies to understand the behavior of the whole power production system. The equivalence ratio was computed as the ratio between the actual oxygen added to the gasifier and the stoichiometric oxygen needed for the complete combustion of the biomass. The gasifier airflow intake was estimated by the following empirical expression [44]:

$$Q_{air\ in}\left(\frac{m^3}{h}\right) = 2.4207 \times (vacuum\ pressure(\ in\ H_2O))^{0.5227} \tag{29}$$

The resulting equivalence ratios were between 0.20 and 0.23. The lower heating value (LHV) of the dry gas was computed based on the molar fractions of fuel gases (Y) and the corresponding LHV at reference conditions [45],

$$LHV_{gas} = 10.79Y_{H_2} + 12.62Y_{CO} + 35.81Y_{CH_4} + 56.08\ Y_{C_2H_2} + 59.04\ Y_{C_2H_4} + 63.75\ Y_{C_2H_6} \tag{30}$$

The LHV of the dry gas was found to be between 5.8 and 6.6 MJ/Nm^3, with the higher values obtained for equivalence ratios of 0.20. The gas yield was estimated based on the mass balance of N_2 in the reactor. It was assumed that all the nitrogen in the fuel exits in the produced gas as N_2 and the N_2 behaves as an inert gas. The gas yields values obtained for the gasification experiments were between 2.05 and 2.20 Nm^{3}gas/kg BSG.

Carbon conversion efficiency (CCE) defines the fraction of solid carbon converted to gas carbon in the produced gas stream. It is clearly a measure of the amount of unconverted carbon and furnishes an indication of the chemical efficiency of the process. Values of CCE between 85.9 and 87.8% were obtained for the experimental conditions used.

Cold gas efficiency (CGE) was computed as the ratio of the chemical energy in the produced gas and the chemical energy in the biomass. Values of CGE between 74.6 and 82.5% were obtained for the experimental conditions used.

The total efficiency is calculated based on the ratio between the power produced (P_{el}) and the chemical energy in the biomass as follows:

$$\eta_t = \frac{P_{el} \times 3.6}{\dot{m}_b \times LHV_b} \tag{31}$$

Values of total efficiency between 15.8% and 17.7% were obtained for the experimental conditions used, which are consistent with other published works [44].

Table 5. Experimental operating conditions and producer gas analysis for brewer's spent grains (BSG).

Experimental Conditions	Run 1	Run 2	Run 3	Run 4	Run 5	Run 6
Biomass feed rate (kg/h)	4.0	3.9	3.7	6.9	6.8	6.5
Air feed rate (Nm^3/h)	4.9	4.3	4.0	8.1	7.7	7.0
Equivalence ratio	0.23	0.21	0.20	0.22	0.21	0.20
Produced gas fraction (%vol. db)						
H_2	14.4 ± 1.7	15.9 ± 0.5	16.4 ± 0.4	15.6 ± 0.9	15.7 ± 0.4	16.6 ± 0.2
CO	15.9 ± 1.5	16.0 ± 0.4	16.2 ± 0.2	16.3 ± 0.2	16.8 ± 0.5	16.9 ± 0.5
CH_4	3.2 ± 1.4	3.4 ± 0.9	3.5 ± 0.9	2.5 ± 0.7	2.5 ± 0.6	2.9 ± 0.4
CO_2	16.3 ± 1.8	16.2 ± 0.7	15.6 ± 0.3	15.8 ± 0.3	15.1 ± 0.6	15.1 ± 0.7
N_2	46.2 ± 2.1	43.8 ± 2.6	42.7 ± 3.4	46.8 ± 1.8	44.9 ± 2.0	43.2 ± 2.9
O_2	2.2 ± 0.3	2.4 ± 0.6	3.2 ± 0.6	2.1 ± 0.4	2.2 ± 0.5	2.5 ± 0.9
C_2H_2	0.0 ± 0.0	0.0 ± 0.0	0.1 ± 0.0	0.0 ± 0.0	0.0 ± 0.0	0.0 ± 0.0
C_2H_4	es1.5 ± 0.6	2.0 ± 0.1	2.0 ± 0.1	2.1 ± 0.1	2.4 ± 0.3	2.5 ± 0.3
C_2H_6	0.3 ± 0.0	0.3 ± 0.0	0.3 ± 0.0	0.3 ± 0.0	0.3 ± 0.0	0.3 ± 0.0
Gasification process characteristics						
Gas LHV (MJ/Nm^3)	5.8	6.3	6.5	5.9	6.4	6.6
Gas yield (Nm^3)	8.5	8.0	7.9	15.2	14.2	13.4
Cold gas efficiency (%)	74.6	78.7	82.4	79.5	80.6	82.5
Carbon conversion efficiency (%)	87.8	86.3	87.6	86.3	86.1	85.9
Power output characteristics						
Power output (kWh)	3.0	3.0	3.0	5.0	5.0	5.0
Total efficiency (%)	16.4	16.8	17.7	15.8	16.0	16.8

3.2. Model Validation

To validate the developed modified stoichiometric equilibrium model, the numerical results were compared with the experimental results obtained in this work. Figure 3 shows a comparison between the numerical results predicted by the developed model (shown in the horizontal axis) and the experimental data (shown in the vertical axis).

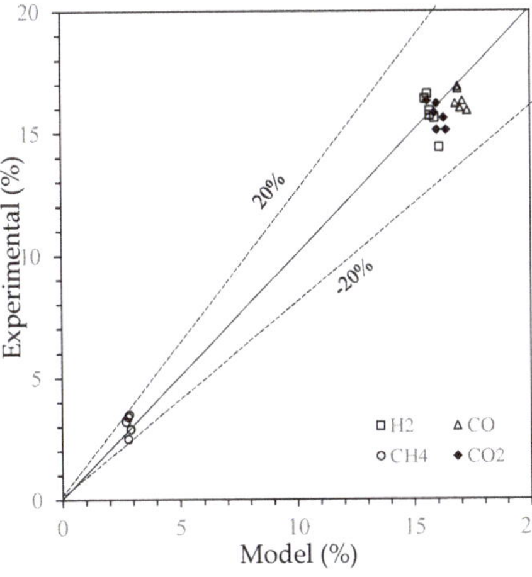

Figure 3. Comparison between modeled and measured produced gas composition for BSG gasification.

From Figure 3, we can see that the developed modified stoichiometric equilibrium model is capable of predicting the produced gas composition within a margin of error of less than 20%. This is a very satisfactory performance for a complex process such as biomass gasification. Greater divergences were detected for methane since reduced molar fractions tend to yield greater relative errors. Moreover, all light hydrocarbons and tars not considered in the model can be lumped into CH_4, which can further explain the higher deviation [46]. Moreover, some degree of divergence should be attributed to the model's nature and assumptions.

3.3. Syngas Quality Assessment for Various Biomass Substrates

Forest and agriculture residues are the most common biomass resources available in Portugal [2, 20,47]. Pine, eucalyptus, and cork are the species representative of the forest residues that are most abundant in Portugal [20,47]. Regarding agricultural residues, the most common are the ones deriving from the agricultural activities of the olive oil and wine industries [20]. These biomass resources can be utilized on a larger scale for energy production and were characterized to evaluate their potential using the developed modified stoichiometric equilibrium model. The characterization of selected biomasses is presented in Table 6.

The produced gas quality depends on the ER, which should be substantially lower than stoichiometry to guarantee that the biomass is gasified instead of burned [48]. An excessively low ER results in frequent problems, such as incomplete gasification and minor LHV of the produced gas. A high ER results in excessive formation of combustion products at the expense of fuel gases [48]. According to Narvaez et al. [49], the ER optimum range for biomass gasification lies between 0.2 and 0.4, and this ER interval was therefore used in our analysis. Regarding the gasifying agent, only air is used in our assessment. The reason lies in the fact of air being the most commonly used gasifying agent, as it is obviously economical [50], and it generates a produced gas of low calorific value, due mainly to its high nitrogen content [51]. Steam as a gasifying agent generates a produced gas with a moderate heating value, and its costs are between air and oxygen. Oxygen is the most expensive gasifying agent and, therefore, used only for more advanced applications [4]. Other operating parameters such as pressure or catalysts use can have a great influence on the produced gas quality [19,26]. However, these are beyond the scope of the present assessment.

Table 6. Characterization of selected biomasses.

Biomass Properties	Pine	Eucalyptus	Cork	Olive Bagasse	Vine Pruning
Proximate analysis (%, ar)					
Ash	2.6	7.5	9.4	13.1	2.7
Volatile	53.6	41.7	62.1	57.8	72.5
Fixed carbon	36.4	44.0	13.4	19.7	11.5
Moisture	7.4	6.8	15.1	9.4	13.3
Ultimate analysis (%, daf)					
C	42.7	43.7	45.2	43.2	41.3
H	5.8	5.6	5.3	5.6	5.5
N	2.1	0.5	0.6	1.9	2.6
S	0.0	0.0	0.0	0.0	0.0
O	49.4	50.2	48.9	49.3	50.6
HHV (MJ/kg)	18.4	17.8	16.4	17.5	15.1

The results obtained using the developed modified stoichiometric equilibrium model of the downdraft gasification using air as the gasifying agent are presented in Table 7 as a function of ER for the various Portuguese biomasses.

Table 7. Model results for various Portuguese biomasses.

Biomass	Pine		
Simulation conditions (ER)	0.20	0.30	0.40
Produced gas fraction (%vol. db)			
H_2	18.0	16.0	14.3
CO	19.0	17.5	16.8
CH_4	2.9	3.0	3.1
CO_2	12.0	13.5	14.7
Biomass	**Eucalyptus**		
Simulation conditions (ER)	0.20	0.30	0.40
Produced gas fraction (%vol. db):			
H_2	16.0	14.8	13.7
CO	18.0	16.6	15.8
CH_4	2.1	2.3	2.4
CO_2	14.0	15.0	15.9
Biomass	**Cork**		
Simulation conditions (ER)	0.20	0.30	0.40
Produced gas fraction (%vol. db)			
H_2	17.5	15.8	14.0
CO	19.2	17.6	16.9
CH_4	3.0	3.1	3.2
CO_2	11.8	13.3	14.5
Biomass	**Olive Bagasse**		
Simulation conditions (ER)	0.20	0.30	0.40
Produced gas fraction (%vol. db)			
H_2	16.0	14.8	13.7
CO	18.3	16.7	15.9
CH_4	2.1	2.3	2.4
CO_2	14.5	15.4	16.1
Biomass	**Vine Prunings**		
Simulation conditions (ER)	0.20	0.30	0.40
Produced gas fraction (%vol. db)			
H_2	21.0	19.9	18.7
CO	20.0	18.4	17.5
CH_4	1.9	2.0	2.1
CO_2	11.0	11.7	12.5

Applications for produced gas can be divided into two main groups—power or heat and fuels or chemical products. Table 8 recapitulates required produced gas characteristics for various end-use options.

Table 8. Produced gas characteristic guidelines for different applications [52].

Application	H_2/CO	Hydrocarbons	N_2	CO_2	Heating Value
Synthetic fuels	0.6	Low	Low	Low	Irrelevant
Methanol	2.0	Low	Low	Low	Irrelevant
Hydrogen	High	Low	Low	Not critical	Irrelevant
Boiler	Irrelevant	High	Irrelevant	Not critical	High
Turbine	Irrelevant	High	Irrelevant	Not critical	High

Typically, produced gas characteristics are more important for chemicals and fuel synthesis applications than for hydrogen and fuel gas applications. Some process equipment such as scrubbers and coolers can be utilized to correct the characteristics of the produced gas to match those ideals for the chosen end-use. However, this supporting equipment increases the complexity and final cost of the process [52].

Figure 4 shows the influence of ER on the produced gas H_2/CO molar ratio for the various biomasses of Table 6.

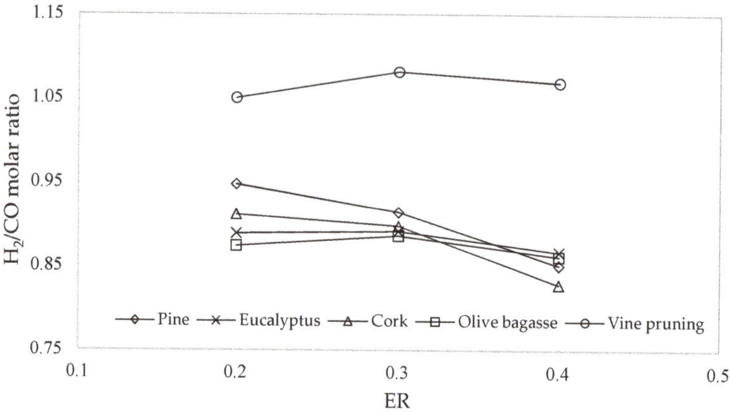

Figure 4. Produced gas H_2/CO molar ratio as a function of equivalence ratio (ER) for various biomasses.

From analyses of Figure 4, it can be seen similar behavior of the H_2/CO molar ratio for most of the biomasses under study. The exception is verified for vine pruning, which presents greater H_2/CO molar ratios. The explanation is linked to biomasses composition provided by proximate and ultimate analyses of Table 6. From Table 6, it is possible to verify the similar composition of the various biomasses being the distinctive aspect of the greater percentage of volatiles of vine pruning. The volatiles are released in the pyrolysis phase, generating CO, H_2, and hydrocarbons as pyrolytic gas products [53]. On the other hand, the increase of ER implies the supply of greater amount of air to the reactor, which favors the oxidation reactions [54]. A low ER ensures high produced gas quality due to higher values of the combustible gases. However, the ER should not be too low because the oxygen supply will not be enough to convert the char. Figure 5 shows the effect of ER on the produced gas CH_4/H_2 molar ratio for the various biomasses of Table 6.

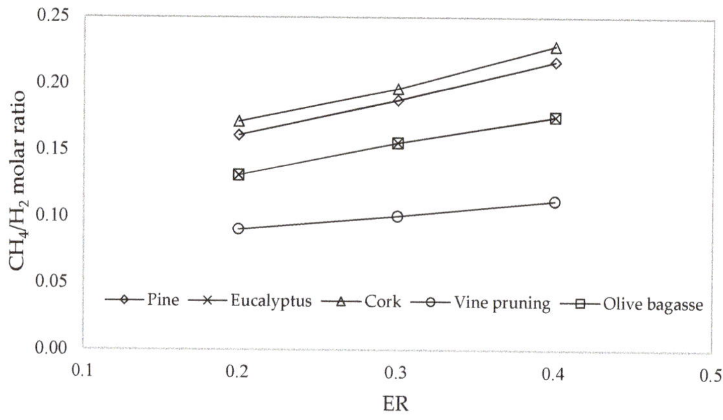

Figure 5. Produced gas CH_4/H_2 molar ratio as a function of ER for various biomasses.

From Figure 5, it is clear that the CH_4/H_2 molar ratio increases with ER for the various biomasses. This behavior is explained by the reducing amounts of H_2 and the approximately constant amounts of CH_4 when increasing ER. It is also seen the similar CH_4/H_2 molar ratios for eucalyptus and olive bagasse. The reason is the similar ultimate composition of those biomasses, as seen in Table 6. The effect of ER on CH_4/H_2 molar ratio is very small since it decreases the H_2 molar fraction that is much greater than the CH_4 molar fraction. In turn, CH_4 molar fraction remains almost constant with the increase of ER from 0.2 to 0.4. From the molar fractions obtained in Table 7 and H_2/CO molar ratio expressed by Figure 4, and according to Table 8, it is possible to conclude that, using air as a gasifying agent, the biomass gasification only provides a syngas with enough quality to be used for energy production in boilers or turbines. The CH_4/H_2 molar ratio expressed in Figure 5, and not directly included in Table 8, is important for domestic purposes [55]. However, the low molar ratios obtained do not identify those produced gases as good candidates for replacing natural gas in domestic applications.

4. Conclusions

In this work, experimental and modeling analysis of brewers' spent grains gasification in a pilot-scale downdraft reactor were performed.

For the experimental part of the work, a pilot-scale integrated gasification power production system called the Power Pallet of 20 kW was used. The gasification process performance was assessed through the produced gas yield and composition, lower heating value, carbon conversion efficiency, and cold gas efficiency. Encouraging results were obtained for all the gasification parameters. The produced gas yields between 2.05 and 2.20 m^3/kg, with composition in between 42.7–46.8 % of N_2, 15.9–16.9% of CO, 15.1–16.3% of CO_2, 14.4–16.6% of H_2, 2.5–3.5% of CH_4, 1.5–2.5% of C_2H_4, and 0.3% of C_2H_6. The heating value of the produced gas was found between 5.8 and 6.6 MJ/m^3, carbon conversion efficiency between 85.9 and 87.8%, and cold gas efficiency between 74.6 and 82.5%. It was also found that the Power Pallet works at low equivalence ratios between 0.20 to 0.23 for the power outputs of 3–5 kWh. Moreover, about 1 kWh electrical power was achieved for approximately 1.3 kg of brewers' spent grains pellets gasified, with an average electrical efficiency of 16.5%. These results are closely in agreement with the power pallet supplier indicative performances.

For the numerical part of the work, a modified stoichiometric equilibrium model of the downdraft gasification was developed to assess the potential applications of the main Portuguese biomasses through produced gas quality indices. The model was validated against the experimental results obtained in the first part of the paper. The sensitivity analysis of the variation of equivalence ratio showed an opposite behavior of the H_2/CO and CH_4/H_2 molar ratios for the biomasses under study. The H_2/CO molar ratio decreases with ER, and the CH_4/H_2 molar ratio decreases with ER. The reason is the similar ultimate composition of the biomasses under study. The exception is verified for vine prunings, which present greater H_2/CO molar ratios when ER increases. The explanation is on the distinctive aspect of the vine pruning proximate composition, which shows a greater percentage of volatiles that is released in the pyrolysis stage generating H_2 and CO.

A final conclusion could be drawn that using air as a gasifying agent in the biomass gasification only provides a produced gas with enough quality to be used for energy production in boilers or turbines. Even for domestic purposes, it is not a good candidate for replacing natural gas.

Author Contributions: Conceptualization, S.F. and E.M.; methodology, S.F., P.B. and C.V.; software, V.S. and S.F.; validation, S.F., E.M. and P.B.; investigation, S.F., L.C. and V.S.; resources, L.C. and V.S.; data curation, S.F. and E.M.; writing—original draft preparation, S.F.; writing—review and editing, E.M., P.B. and C.V.; supervision, E.M., P.B. and C.V.

Funding: This research was funded by the FOUNDATION FOR SCIENCE AND TECHNOLOGY (FCT), grant number SFRH/BD/91894/2012.

Conflicts of Interest: The authors declare no conflict of interest.

References

1. IEA. *World Energy Outlook 2018*; IEA: Paris, France, 2018. [CrossRef]
2. Ferreira, S.; Moreira, N.A.; Monteiro, E. Bioenergy overview for Portugal. *Biomass Bioenergy* **2009**, *33*, 1567–1576. [CrossRef]
3. Faaij, A.P.C. Bio-energy in Europe: Changing technology choices. *Energy Policy* **2006**, *34*, 322–342. [CrossRef]
4. Kirkels, A.; Verbong, G. Biomass Gasification: Still promising? A 30-year global overview. *Renew. Sust. Energy Rev.* **2011**, *15*, 471–481. [CrossRef]
5. Ahmad, A.A.; Zawawi, N.A.; Kasim, F.H.; Inayat, A.; Khasri, A. Assessing the gasification performance of biomass: A review on biomass gasification process conditions, optimization and economic evaluation. *Renew. Sust. Energy Rev.* **2016**, *53*, 1333–1347. [CrossRef]
6. Prasad, S.; Singh, A.; Joshi, H.C. Ethanol as an alternative fuel from agricultural, industrial and urban residues. *Resour. Conserv. Recycl.* **2007**, *50*, 1–39. [CrossRef]
7. Liguori, R.; Amore, A.; Faraco, V. Waste valorization by biotechnological conversion into added value products. *Appl. Microbiol. Biotechnol.* **2013**, *97*, 6129–6147. [CrossRef]
8. Lynch, K.M.; Steffen, E.J.; Arendt, E.K. Brewers' spent grain: A review with an emphasis on food and health. *J. Inst. Brew.* **2016**, *122*, 553–568. [CrossRef]
9. Mussatto, S.I. Brewer's spent grain: A valuable feedstock for industrial applications. *Sci. Food Agric.* **2014**, *94*, 1264–1275. [CrossRef]
10. Mussatto, S.I.; Dragone, G.; Roberto, I.C. Brewers' spent grain: Generation, characteristics and potential applications. *J. Cereal Sci.* **2006**, *43*, 1–14. [CrossRef]
11. Ulbricha, M.; Preßla, D.; Fendta, S.; Gadererc, M.; Spliethoff, H. Impact of HTC reaction conditions on the hydrochar properties and CO_2 gasification properties of spent grains. *Fuel Proc. Technol.* **2017**, *167*, 663–669. [CrossRef]
12. Aliyu, S.; Bala, M. Brewer's spent grain: A review of its potentials and applications. *Afr. J. Biotechnol.* **2011**, *10*, 324–331. [CrossRef]
13. Ferreira, S.; Monteiro, E.; Brito, P.; Costa, C.; Calado, L.; Vilarinho, C. Experimental analysis of beer spent grains steam gasification in an allothermal batch reactor. *Energies* **2019**, *12*, 912. [CrossRef]
14. Mathias, T.R.S.; Mello, P.P.M.; Servulo, E.F.C. Solid wastes in brewing process: A review. *J. Brew. Distill.* **2014**, *5*, 1–9. [CrossRef]
15. Stewart, G.G.; Russel, I.; Anstruther, A. *Handbook of Brewing*, 3rd ed.; CRC Press: Boca Raton, IL, USA, 2017; pp. 83–538. [CrossRef]
16. Briggs, D.E.; Boulton, C.A.; Brookes, P.A.; Stevens, R. *Brewing Science and Practice*, 1st ed.; Woodhead Publishing Limited: Cambridge, UK, 2004. [CrossRef]
17. Kerby, C.; Vriesekoop, F. An Overview of the Utilisation of Brewery By-Products as Generated by British Craft Breweries. *Beverages* **2017**, *3*, 24. [CrossRef]
18. O'Rourke, T. Making the most of your hops. *New Brewer.* **1994**, *11*, 20–33.
19. Couto, N.; Rouboa, A.; Silva, V.; Monteiro, E.; Bouziane, K. Influence of the biomass gasification processes on the final composition of syngas. *Energy Procedia* **2013**, *36*, 596–606. [CrossRef]
20. Ferreira, S.; Monteiro, E.; Brito, P.; Vilarinho, C. Biomass resources in Portugal: Current status and prospects. *Renew. Sust. Energy Rev.* **2017**, *78*, 1221–1235. [CrossRef]
21. Mahmood, A.S.N.; Brammer, J.G.; Hornung, A.; Steele, A.; Poulston, S. The intermediate pyrolysis and catalytic steam reforming of Brewers spent grain. *J. Anal. Appl. Pyrolysis* **2013**, *103*, 328–342. [CrossRef]
22. Borel, L.D.M.S.; Lira, T.S.; Ribeiro, J.A.; Ataíde, C.H.; Barrozo, M.A.S. Pyrolysis of brewer's spent grain: Kinetic study and products identification. *Ind. Crop. Prod.* **2018**, *21*, 388–395. [CrossRef]
23. Couto, N.; Silva, V.; Monteiro, E.; Brito, P.; Rouboa, A. Modeling of fluidized bed gasification: Assessment of zero-dimensional and CFD approaches. *J. Sci.* **2015**, *24*, 378–385. [CrossRef]
24. Ramos, A.; Monteiro, E.; Rouboa, A. Numerical approaches and comprehensive models for gasification process: A review. *Renew. Sust. Energy Rev.* **2019**, *110*, 188–206. [CrossRef]
25. Ferreira, S.; Monteiro, E.; Brito, P.; Vilarinho, C. A holistic review on biomass gasification modified equilibrium models. *Energies* **2019**, *12*, 160. [CrossRef]
26. Basu, P. *Biomass Gasification and Pyrolysis: Practical Design and Theory*, 2nd ed.; Elsevier: Oxford, UK, 2013; ISBN 978-0-12-374988-8.

27. La Villetta, M.; Costa, M.; Massarotti, N. Modelling approaches to biomass gasification: A review with emphasis on the stoichiometric method. *Renew. Sust. Energy Rev.* **2017**, *74*, 71–88. [CrossRef]

28. Altafini, C.R.; Wander, P.R.; Barreto, R.M. Prediction of the working parameters of a wood waste gasifier through an equilibrium model. *Energy Convers. Manag.* **2003**, *44*, 2763–2777. [CrossRef]

29. Zainal, Z.A.; Ali, R.; Lean, C.H.; Seetharamu, K.N. Prediction of performance of a downdraft gasifier using equilibrium modeling for different biomass materials. *Energy Convers. Manag.* **2001**, *42*, 1499–1515. [CrossRef]

30. Azzone, E.; Morini, M.; Pinelli, M. Development of an equilibrium model for the simulation of thermochemical gasification and application to agricultural residues. *Renew. Energy* **2012**, *46*, 248–254. [CrossRef]

31. Silva, V.; Rouboa, A. Using a two-stage equilibrium model to simulate oxygen air enriched gasification of pine biomass residues. *Fuel Process. Technol.* **2013**, *109*, 111–117. [CrossRef]

32. ALL Power Labs. "GEK Gasifier". 2013. Available online: http://www.gekgasifier.com/ (accessed on 9 February 2019).

33. Monteiro, E.; Sotton, J.; Bellenoue, M.; Moreira, N.A.; Malheiro, S. Experimental study of syngas combustion at engine-like conditions in a rapid compression machine. *Exp. Fluid Sci.* **2011**, *35*, 1473–1479. [CrossRef]

34. Allesina, G.; Pedrazzi, S.; Allegretti, F.; Morselli, N.; Puglia, M.; Santunione, G.; Tartarini, P. Gasification of cotton crop residues for combined power and biochar production in Mozambique. *Appl. Eng.* **2018**, *139*, 387–394. [CrossRef]

35. Ayol, A.; Yurdakos, O.T.; Gurgen, A. Investigation of municipal sludge gasification potential: Gasification characteristics of dried sludge in a pilot-scale downdraft fixed bed gasifier. *Int. J. Hydrog. Energy* **2019**, *44*, 17397–17410. [CrossRef]

36. Prins, M.J.; Ptasinski, K.J.; Janssen, F.J.J.G. Thermodynamics of gas-char reactions first and second law analysis. *Chem. Eng. Sci.* **2003**, *58*, 1003–1011. [CrossRef]

37. Ptasinski, K.J.; Prins, M.J.; Pierik, A. Exergetic evaluation of biomass gasification. *Energy* **2007**, *32*, 568–574. [CrossRef]

38. Karamarkovic, R.; Karamarkovic, V. Energy and exergy analysis of biomass gasification at different temperatures. *Energy* **2010**, *35*, 537–549. [CrossRef]

39. Jarungthammachote, S.; Dutta, A. Thermodynamic equilibrium model and second law analysis of a downdraft waste gasifier. *Energy* **2007**, *32*, 1660–1669. [CrossRef]

40. Probstein, R.F.; Hicks, R.E. *Synthetic Fuel*; McGraw-Hill: New York, NY, USA, 1982.

41. De Souza-Santos, M.L. *Solid Fuels Combustion and Gasification: Modeling, Simulation, and Equipment Operation*; Marcel Dekker: New York, NY, USA, 2004.

42. Channiwala, S.A.; Parikh, P.P. A unified correlation for estimating HHV of solid, liquid and gaseous fuels. *Fuel* **2002**, *81*, 1051–1063. [CrossRef]

43. Puig-Arnavat, M.; Bruno, J.C.; Coronas, A. Review and analysis of biomass gasification models. *Renew. Sust. Energy Rev.* **2010**, *14*, 2841–2851. [CrossRef]

44. Dion, L.-M.; Lefsrud, M.; Orsat, V.; Cimon, C. Biomass gasification and syngas combustion for greenhouse CO_2 enrichment. *Bioresources* **2013**, *8*, 1520–1538. [CrossRef]

45. Kaewluan, S.; Pipatmanomai, S. Potential of synthesis gas production from rubber wood chip gasification in a bubbling fluidised bed gasifier. *Energy Convers. Manag.* **2011**, *52*, 75–84. [CrossRef]

46. Couto, N.; Silva, V.; Monteiro, E.; Rouboa, A. Exergy analysis of Portuguese municipal solid waste treatment via steam gasification. *Energy Convers. Manag.* **2017**, *134*, 235–246. [CrossRef]

47. Monteiro, E.; Mantha, V.; Rouboa, A. The Feasibility of Biomass Pellets Production in Portugal. *Energy Source. Part B* **2013**, *8*, 28–34. [CrossRef]

48. Monteiro, E.; Ismail, T.M.; Ramos, A.; Abd El-Salam, M.; Brito, P.; Rouboa, A. Assessment of the miscanthus gasification in a semi-industrial gasifier using a CFD model. *Appl. Eng.* **2017**, *123*, 448–457. [CrossRef]

49. Narvaez, I.; Orío, A.; Aznar, M.P.; Corella, J. Biomass gasification with air in an atmospheric bubbling fluidized bed. Effect of six operational variables on the quality of the produced raw gas. *Ind. Eng. Chem. Res.* **1996**, *35*, 2110–2120. [CrossRef]

50. Ruiz, J.A.; Juárez, M.C.; Morales, M.P.; Muñoz, P.; Mendívil, M.A. Biomass gasification for electricity generation: Review of current technology barriers. *Renew. Sust. Energy Rev.* **2013**, *18*, 174–183. [CrossRef]

51. Wang, L.; Weller, C.L.; Jones, D.D.; Hanna, M.A. Contemporary issues in thermal gasification of biomass and its application to electricity and fuel production. *Biomass Bioenergy* **2008**, *32*, 573–581. [CrossRef]

52. Ciferno, J.P.; Marano, J.J. *Benchmarking Biomass Gasification Technologies for Fuels, Chemicals and Hydrogen Production*; U.S. Department of Energy, National Energy Technology Laboratory: Pittsburgh, PA, USA, 2002.
53. Hua, X.; Gholizadeh, M. Biomass pyrolysis: A review of the process development and challenges from initial researches up to the commercialisation stage. *J. Energy Chem.* **2019**, *39*, 109–143. [CrossRef]
54. Silva, V.; Couto, N.; Monteiro, E.; Rouboa, A. Assessment of municipal solid wastes gasification in a semi-industrial gasifier using syngas quality indices. *Energy* **2015**, *93*, 864–873. [CrossRef]
55. Silva, V.; Monteiro, E.; Couto, N.; Brito, P.; Rouboa, A. Analysis of syngas quality from Portuguese biomasses: An experimental and numerical study. *Energy Fuels* **2014**, *28*, 5766–5777. [CrossRef]

Article

Techno-Economic Assessment of the Use of Syngas Generated from Biomass to Feed an Internal Combustion Engine

J. R. Copa [1], C. E. Tuna [1], J. L. Silveira [1], R. A. M. Boloy [2], P. Brito [3], V. Silva [3,4,*], J. Cardoso [3,5] and D. Eusébio [3]

1   Faculty of Engineering, Campus of Guaratingueta, Bioenergy Research Institute (IPBEN), Laboratory of Energy Systems Optimization (LOSE), Av. Dr. Ariberto Pereira da Cunha, Sao Paulo State University, 333 Portal das Colinas Guaratingueta, SP 12516-410, Brazil; jcoparey@gmail.com (J.R.C.); celso.tuna@unesp.br (C.E.T.); jose.luz@unesp.br (J.L.S.)
2   Group of Entrepreneurship, Energy, Environment and Technology-GEEMAT, Mechanical Engineering Department, Federal Center of Technological Education of Rio de Janeiro-CEFET/RJ, Av. Maracanã, 229, Rio de Janeiro, RJ 20271-110, Brazil; ronney.boloy@cefet-rj.br
3   VALORIZA, Polytechnic Institute of Portalegre, 7300-555 Portalegre, Portugal; pbrito@ipportalegre.pt (P.B.); jps.cardoso@ipportalegre.pt (J.C.); danielafle@ipportalegre.pt (D.E.)
4   ForestWise, Collaborative Laboratory for Integrated Forest & Fire Management, 5001-801 Vila Real, Portugal
5   Instituto Superior Técnico, Universidade de Lisboa, 1649-004 Lisboa, Portugal
*   Correspondence: valter.silva@ipportalegre.pt or valter.silva@forestwise.pt; Tel.: +351-245-301-592

Received: 30 April 2020; Accepted: 11 June 2020; Published: 15 June 2020

Abstract: The focus of this study is to provide a comparative techno-economic analysis concerning the deployment of small-scale gasification systems in dealing with various fuels from two countries, Portugal and Brazil, for electricity generation in a 15 kWe downdraft gasifier. To quantify this, a mathematical model was implemented and validated against experimental runs gathered from the downdraft reactor. Further, a spreadsheet economic model was developed combining the net present value (NPV), internal rate of return (IRR) and the payback period (PBP) over the project's lifetime set to 25 years. Cost factors included expenses related to electricity generation, initial investment, operation and maintenance and fuel costs. Revenues were estimated from the electricity sales to the grid. A Monte Carlo sensitivity analysis was used to measure the performance of the economic model and determine the investment risk. The analysis showed an electricity production between 11.6 to 15 kW, with a general system efficiency of approximately 13.5%. The viability of the projects was predicted for an NPV set between 18.99 to 31.65 k€, an IRR between 16.88 to 20.09% and a PBP between 8.67 to 12.61 years. The risk assessment yielded favorable investment projections with greater risk of investment loss in the NPV and the lowest for IRR. Despite the feasibility of the project, the economic performance proved to be highly reliant on the electricity sales prices subdue of energy market uncertainties. Also, regardless of the broad benefits delivered by these systems, their viability is still strikingly influenced by governmental decisions, subsidiary support and favorable electricity sales prices. Overall, this study highlights the empowering effect of small-scale gasification systems settled in decentralized communities for electric power generation.

Keywords: biomass gasification; internal combustion engines-generator; small-scale systems; energy efficiency; techno-economic analysis; Monte Carlo method

1. Introduction

The shortage and unpredictability of conventional energy sources affected by depletion and global geopolitical issues are causing an energy crisis that is accelerating the renewable energy use [1,2].

In Portugal, 28.6% of the consumed energy derives from renewable energy sources (RES), with wind accounting for 7.2%, hydroelectric 7.4%, solar 1.0%, geothermal 0.1%, and, the most preeminent, bioenergy, representing 12.9% of the total [3]. As for Brazil, the power generation from RES reaches 43.2%, out of which hydroelectric represents 11.9%, bioenergy 25.4% and other renewables combined (solar, wind and geothermal) 5.9% [4]. In both countries, the power generation from RES is above the world average, which is approximately 13.5% [5]. However, these RES may not always be available when required. Biomass products, if properly managed, can be collected and used to produce energy regardless of environmental conditions. The exploration for energy purposes of different biomass or their mixtures could increase the generation of energy and contribute to reducing fossil fuels consumption [6].

One interesting route to convert biomass and solid waste into energy is through gasification systems. When compared to other technologies such as combustion, gasification provides enhanced efficiency and environmental performance, meeting the ever-increasing environmental restrictions imposed by governments and international agencies [6]. Gasification can be best defined as the conversion of biomass and/or solid waste to syngas by oxidation of the feedstock under fuel-rich combustion conditions [7].

Syngas' properties allow it to be burned in standard spark-ignition (SI) and compression ignition (CI) engines, however, the lower energy density from the syngas/air mixture reduces maximum brake power significantly [8]. A promising syngas application is its use in internal combustion engines-generator (ICEG) [9]. The use of such engines presents several advantages over turbines and steam generators such as bearing a wide range of power ratings, being capable of running on different fuels, and also the strong know-how concerning its system management and maintenance [9].

The replacement of diesel fuel with syngas is highly beneficial for soot emissions (due in part to the combined effects of soot formation and oxidation rates inside the cylinder when syngas is present) and also the power rating of an engine running on syngas/diesel is less affected than in an engine running on syngas/gasoline, arising high expectations concerning its use [10,11].

Particle and tar concentration in the syngas must be less than 50 mg/Nm^3 and 100 mg/Nm^3, respectively, for the satisfactory operation of an ICEG [12]. Under these circumstances, choosing to operate with a downdraft gasifier seems to be the most fitting solution among available gasification technologies [9]. This is due to its inherent low particulate and tar content rate which relates to the internal design of these systems as the tar produced throughout the pyrolysis stage gets thermally converted into gas during the combustion stage of the gasification process [9].

Several studies have been reported on power generation plants based on downdraft gasification and ICEG. Mancebo et al. [13], presented an analysis of a cogeneration plant composed of a downdraft gasifier coupled to an ICEG with a capacity of 15 kWe, using eucalyptus biomass as fuel. The authors declared an electricity generation efficiency of 12.5% regarding a biomass consumption of 10 kg/h.

Dasappa et al. [14] reported the experimental operation of a 100 kWe gasification power plant using wood as feedstock. Here, a downdraft reactor was employed with the capacity to process 110 kg/h of woody biomass. The specific biomass consumption was 1.36 kg/kWh and the electricity generation efficiency of 18%.

Lee et al. [15] performed an experimental evaluation of four biomasses in a generation system consisting of a downdraft gasifier and an SI engine for electric power generation. Results showed that the general efficiency of the system varied from 15.8 to 23.0% depending on the biomass used, and the electric power produced varied between 10.1 and 13.1 kWe.

Elsner et al. [16] presented a downdraft gasification system coupled with an ICEG and a waste heat recovery system for the gasification of waste sludge. Results indicated that due to the low LHV of the syngas, it is recommended to mix wood pellets with sewage sludge (40/60%). Economic results also showed that due to current energy market conditions, it is more appropriate to use the generated heat and electricity for self-consumption.

Although downdraft gasification technology is well known and studied, more experimental studies are still required due to the variability in biomass composition. Experimental tests can be expensive; therefore, numerical methods such as computational fluid dynamics (CFD) emerge as a very powerful and useful tool. The literature provides several works concerning the use of CFD approaches to model gasification process [8,17]. Silva et al. [18], developed a multiphase CFD model able to predict the syngas composition over different operating conditions, gasifying agents and feedstocks using a pilot-scale gasifier. Results showed that syngas properties can be tailored depending on the selected feedstock and operating conditions, and that syngas production can be controlled to minimize composition oscillations using adequate statistical strategies.

Table 1 summarizes a set of results from experimental studies related to the generation of electricity from downdraft gasification with different biomasses.

Table 1. Various experimental studies on electricity generation using downdraft gasification.

Authors	Year	Case Study	Syngas LHV (MJ/Nm^3)	Original Fuel	Max. Power (kW)	Efficiency (%)
Coronado et al., [19]	2011	Analyzes the gasification in a wood downdraft and its integration with a compact cogeneration system for applications in rural communities in Brazil.	5.6	Gasoline; NG	15	21.4 (Elec. eff. ICE)
Luz et al. [20]	2015	Study MSW downdraft gasification for electricity generation from ICE and possible applications in small municipalities in Brazil.	4.6	Gasoline	97	23 (Elec. eff.)
Raman et al. [21]	2013	Analyzes the performance of an ICE fed with syngas generated in a downdraft gasifier	5.6	NG	73	21 (Overall eff.)
Indrawan et al. [22]	2017	Study the gasification of low-density biomass (Switchgrass), in a downdraft gasifier modified and the application of syngas for the generation of electricity in ICE.	6 to 7	NG	5	21.3 (Elec. eff.)
La Villetta et al. [23]	2018	The downdraft gasification of wood chips is analyzed and the use of syngas to feed the cogeneration system, consisting of an ICE and two heat exchangers with capacity for 20 kW electric and 40 kW thermal.	3.7	Gasoline	20	22.1 (Elec. eff.)
Chang et al. [24]	2019	Analyzes the joint generation of electricity and heat from the downdraft gasification of rice hulls in the context of Taiwan	3.0	NG	1150	27.9 (Elec. eff.)

NG: Natural Gas.

To the best of our knowledge, there are very few works concerning the numerical simulation of energy generation systems integrating downdraft gasifier with ICEG. As a matter of fact, studies related to combined systems may be found in the literature, yet these fail to provide a thorough analysis concerning the system efficiency or even the most appropriate syngas composition to feed said system. Thus, studies devoted to gasification-ICEG integrated systems are of utmost importance for the successful implementation of this technology, promoting it as an auspicious and realistic solution with applications in solid waste treatment and electrification while contributing to fossil fuels consumption reduction. Finally, and as one of the main focal points of this work, these systems are deemed as especially attractive for decentralized small-scale (≤1 MW) electricity generation in rural and/or remote communities, particularly in developing countries, bearing alternative electric power solutions to communities where connection to the central grid is economically unfeasible. In fact, these units provide a window of opportunity for achieving global access to electricity being rapidly scalable, environmentally sustainable and tailor-made to local conditions, suiting as key for unlocking a sustainable future while uplifting the local economy in these locations. Moreover, biomass exploration provides a helping hand towards wildfire hazards reduction by promoting forest biomass harvesting and cleaning in over-grown areas [16,25].

In this sense, the purpose of this work is to present a CFD simulation obtaining the syngas from a 15 kWe downdraft gasification system in dealing with four different fuels, including biomass and municipal solid waste (MSW), from two countries, Portugal and Brazil, and then feed it into an ICEG. The simulation results are directly compared and validated against experimental data. A techno-economic evaluation coupled with a Monte Carlo sensitivity analysis is performed to assess the feasibility of the units from an economic standpoint. Finally, the influence of biomass characteristics on the composition of the syngas and the efficiency of the cold gas is studied.

2. Experimental System Description

2.1. Downdraft Gasifier Description

The experimental gasification runs were performed using a Power Pallet 15 kWe gasifier from All Power Labs (Berkeley, CA, USA), which is a combination of an Imbert style downdraft fixed bed reactor with an electric power generator and an electronic control unit. A schematic of the unit is depicted in Figure 1a. The downdraft gasifier (Figure 1b) is composed of a biomass storage hopper, which is simultaneously designed to dry the feedstock through the recirculation of the hot gases produced within the reactor. The biomass is supplied from the top while the air moves downwards, being preheated through contact with the walls of the reactor. Ash collection is carried out in a separate tank at the bottom of the reactor, while the produced syngas passes through a cyclone to remove fine particles. The gas is conducted to the biomass hopper for drying and further filtration. Subsequently, one may collect the gas for analysis or directly inject it into the generator. The condensates resulting from the process are collected in a filter [26].

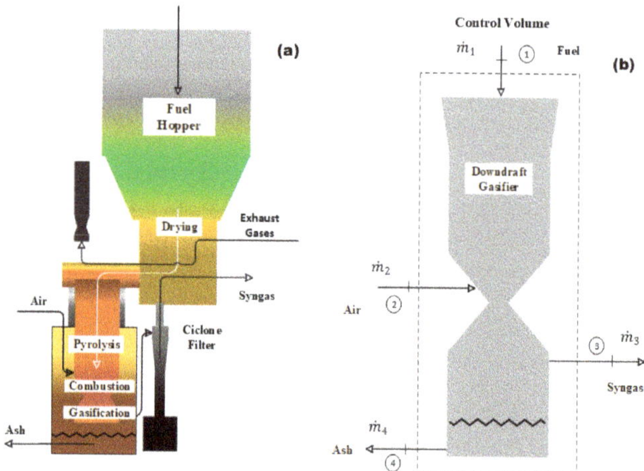

Figure 1. (**a**) Physical diagram of the system, and (**b**) control volume of the downdraft gasifier.

The reactor vessel is cylindrically shaped, with an internal diameter of 28 cm and a height of 55 cm. Inside, a reduction cone tapers to 20 cm and the heart cone tapers to 7 cm so to increase solid residence time in the pyrolysis and combustion zones. The gasifier narrows below the flame zone to restrict the tar content in the syngas by forcing the volatiles to pass through the combustion zone. In the reduction zone, the diameter expands to 19 cm, followed by 23 cm and 38 cm, respectively.

Syngas exits through a 21 cm equivalent diameter outlet. The biomass processing capacity is 22 kg/h, producing 60 m^3/h of syngas [27].

2.2. Compact Generation System

The generation system is composed of an ICEG and an electricity generator. The main characteristics of the generation system are shown in Table 2 [27].

Table 2. Main characteristics of the generation system.

Engine	Generator
• Model: GM Vortec 3.0 L	•Model: Mecc Alte NPE 32
• 4 strokes	• Frequency: 50/60 Hz
• Fuels: Gasoline, LPG, NG	• Number of poles: 4
• Compression value: 9.4:1	• 2250 rpm
• Max power: 37 kW	• Cos φ: 0.8
• Max power torque: 73 Nm	• Power: 25 kVA-20 kW
• Max rotation: 3000 rpm	• 220 V: 32.8 A
LPG: Liquefied Petroleum Gas; NG: Natural Gas.	

The system depicted in Figure 2 is composed of an engine model GM Vortec 3.0 L I-4, coupled to a generator, model NPE 32 (Mecc Alte, Dry Creek, Australia) capable of producing 20 kW of electric power [27].

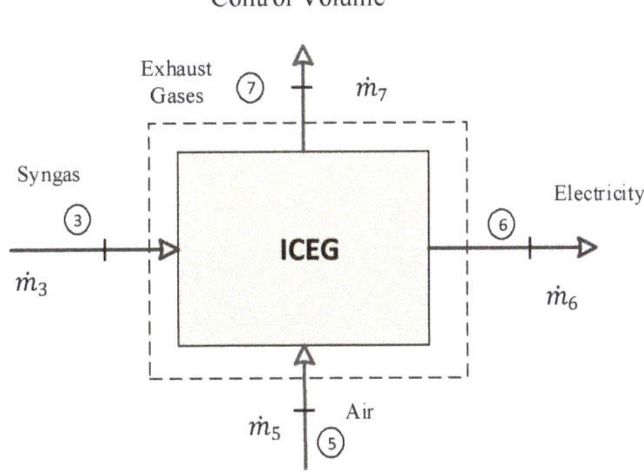

Figure 2. Control volume of the power generation system.

3. Computational Model Set up

3.1. Fuel Characteristics

As the accuracy of the mathematical model relies upon the use of realistic data, biomass and MSW characterization from both Portugal and Brazil were gathered from experimental studies developed by Couto et al. [28], Sales et al. [29] and Luz et al. [20]. Table 3 shows the elemental and proximal analysis for the studied biomasses and MSW from both countries. Throughout this work, MSW * and MSW ** will be used to address MSW coming from Portugal and Brazil, respectively.

Table 3. Chemical composition of the used fuels.

	Portugal		Brazil	
	Acacia Residues	**MSW * [28]**	**Eucalyptus [29]**	**MSW ** [20]**
Elemental Analysis (wt.%, Dry Basis)				
C	44.1	48.0	49.0	49.7
H	5.6	6.3	6.3	7.2
N	0.4	1.4	0.2	0.8
S	0.0	0.7	0.1	0.7
O (by difference)	49.9	43.6	44.4	41.6
Proximal Analysis (wt.%, Wet Basis)				
Moisture	14.2	17.6	11.3	6.5
Ash	4.0	14.9	0.8	14.3
Volatile matter	49.7	76.6	72.7	74.1
Fixed carbon	32.1	8.5	15.2	5.1
LHV (MJ/kg)	17.0	14.4	18.4	19.6

To determine the syngas composition a mathematical model was built gathering the experimental data obtained from the downdraft gasifier. From the model results, the H_2, CO, CH_4, and CO_2 concentrations were obtained and the LHV of the syngas was calculated using Equation (1) [29]:

$$LHV_{Syngas} = 12696 \cdot (CO) + 10768 \cdot (H_2) + 35866 \cdot (CH_4) \tag{1}$$

3.2. Mathematical Model

The computational geometry domain refers to a 2-D downdraft reactor with an internal diameter of 28 cm and a height of 55 cm. The initial ignition of the reactor was carried out with charcoal and then the biomass pellets were introduced, sizing 0.5 cm thick, 1 to 2 cm wide and 2 to 2.5 cm long. The obtained syngas travels between the internal and external walls of the gasifier before entering the cyclone filter to remove particles and then through the full-packed bed filter to eliminate tars. Simulations were performed using a time step size of 0.001 s, for a total number of 50,000 time steps (50 s).

The implemented 2-D Eulerian-Eulerian mathematical model was firstly developed by Silva et al. [30]. Complex phenomena regarding the gasification process for the downdraft reactor were simulated by means of a multiphase (gas and solid) model within the ANSYS Fluent database. The gas-phase was considered as a continuum and solid phase was modeled following a Eulerian granular model. Interactions between phases were modeled as well, with both phases exchanging heat by convection, momentum (due to drag between phases), and mass (given the heterogeneous chemical reactions). To appropriately describe the hydrodynamic phenomena within the fluidized bed reactor the standard k-ε turbulence model is applied. The heat transfer between the solid and gas phases, the viscous dissipation, and the expansion work is described by the energy conservation equation. Table 4 summarizes the main governing equations for both gas and solid phase and the hydrodynamic model. The energy conservation equation describes the heat exchange between gas and solid phases, the viscous dissipation, and the expansion work of the void fraction. The equation for the gas and solid phases is shown in Table 4, in which $\vec{Q}_{pq}$ is the heat transfer intensity, h_q the specific enthalpy of phase, $\vec{q}_q$ the heat flux, S_q the source term and h_{pq} the enthalpy of the interface. For the mass balance model the gas and solid phases continuity equations are provided, the α is the volume fraction, ρ density, v_g velocity and S_{gs} is defined as the source term (Table 4). The momentum equations for both phases consider the gas stress tensor τ_g, gravity g, the gas-solid interface drag coefficient β, and the solid mean velocity U_s (Table 4). The granular Eulerian model treats the continuous fluid (primary phase) as well as dispersed solids (secondary phase) as interpenetrating continua. Stresses in the granular solid phase are obtained by analogy between the random particles motion and the thermal motion of molecules within a gas accounting for the inelasticity of solid particles. The kinetic energy associated with velocity fluctuations is described by granular temperature which is proportional to the norm of particle velocity fluctuations. In the conservation equation for the granular temperature the $\gamma_{\Theta a}$ is the collisional dissipation rate of granular energy, $\vec{v}_s$ the diffusive flux of granular energy, φ_{1s} the granular energy exchange between the gas and solid phase, and $k_{\Theta a}$ is the diffusion coefficient (Table 4). Since the present model has already been extensively documented in recent literature published by the research group, only key points will be highlighted. Further details on the model can be found elsewhere [30].

Table 4. Conservation equations and hydrodynamic model for both gas and solid phases.

Conservation Equations
Energy (gas phase):
$\frac{\partial(\alpha_q \rho_q h_q)}{\partial t} + \nabla(\alpha_q \rho_q \vec{v}_q h_q) = -\alpha_q \frac{\partial(p_q)}{\partial t} + \bar{\tau}_q : \nabla \vec{v}_q - \nabla \vec{q}_q + S_q + \sum_{p=1}^{n}(\vec{Q}_{pq} + \dot{m}_{pq} h_{pq})$
Mass (gas phase):
$\frac{\partial(\alpha_g \rho_g)}{\partial t} + \nabla \cdot (\alpha_g \rho_g v_g) = S_{gs}; \; S_{gs} = -S_{gs} = M_c \sum \gamma_c R_c$
Momentum (gas phase):
$\frac{\partial(\alpha_g \rho_g v_g)}{\partial t} + \nabla \cdot (\alpha_g \rho_g v_g v_g) = -\alpha_g \nabla p_g + \alpha \rho_g g + \beta(v_g - v_s) + \nabla \cdot \alpha_g \tau_g + S_{gs} U_s$
Energy (solid phase):
$\frac{\partial(\alpha_p \rho_p h_p)}{\partial t} + \nabla(\alpha_p \rho_p \vec{v}_p h_p) = -\alpha_p \frac{\partial(p_q)}{\partial t} + \bar{\tau}_p : \nabla \vec{v}_p - \nabla \vec{q}_p + S_p + \sum_{p=1}^{n}(\vec{Q}_{pq} + \dot{m}_{pq} h_{pq})$
Mass (solid phase):
$\frac{\partial(\alpha_s \rho_s)}{\partial t} + \nabla \cdot (\alpha_s \rho_s v_s) = S_{sg}; \; S_{sg} = -S_{sg} = M_c \sum \gamma_c R_c$
Momentum (solid phase):
$\frac{\partial(\alpha_s \rho_s v_s)}{\partial t} + \nabla \cdot (\alpha_s \rho_s v_s v_s) = -\alpha_s \nabla p_s + \alpha \rho_s g + \beta(v_g - v_s) + \nabla \cdot \alpha_s \tau_s + S_{sg} U_s$
Hydrodynamic model
Kinetic energy:
$\frac{\partial}{\partial t}(\rho k) + \frac{\partial}{\partial x_i} \cdot (\rho k u_i) = \frac{\partial}{\partial x_j}\left[(\mu + \frac{\mu_t}{\sigma_\varepsilon})\right] + G_k + G_b - \rho \varepsilon - Y_M + S_k$
Dissipation rate:
$\frac{\partial}{\partial t}(\rho \varepsilon) + \frac{\partial}{\partial x_i} \cdot (\rho \varepsilon u_i) = \frac{\partial}{\partial x_j}\left[(\mu + \frac{\mu_t}{\sigma_\varepsilon}) \frac{\partial \varepsilon}{\partial x_j}\right] + C_{1\varepsilon} \frac{\varepsilon}{k}(G_k + G_{3\varepsilon} G_b) - C_{2\varepsilon} \rho \frac{\varepsilon^2}{k} + S_\varepsilon$
Granular Eulerian model:
$\frac{3}{2}\left[(\frac{\partial(\rho_s \alpha_s \Theta_s)}{\partial t} + \nabla \cdot (\rho_s \alpha_s \vec{v}_s \Theta_s))\right] = (-P_s \bar{I} + \bar{\tau}_s) : \nabla(\vec{v}_s) + \nabla \cdot (k_{\Theta a} \nabla(\Theta_s)) - \gamma_{\Theta a} + \varphi_{1s}$

3.3. Chemical Reactions Model

In the chemical reaction model, the kinetic/diffusion surface reaction model portrays the heterogeneous reactions, and the finite-rate/eddy-dissipation model is used to describe the homogeneous gas-phase reactions. The gasification of biomass and MSW involves several fundamental processes. First, volatile components, such as light gases and tar are released by pyrolysis. These species undergo homogeneous gas-phase reactions forming CO, CO_2, H_2 and H_2O which then combust and gasify the char. Table 5 provides the devolatilization model, main chemical reactions and reaction rates coefficients (based on the Arrhenius law) in the chemical model. Additional information regarding the model can be found elsewhere [30].

Table 5. Implemented chemical reactions model.

Type.	Chemical Reactions and Arrhenius Rate
Pyrolysis	$Cellulose \xrightarrow{r_1} \alpha_1 volatiles + \alpha_2 TAR + \alpha_3 char; \; r_1 = A_t exp\left(\frac{-E_i}{T_s}\right)(1-a_i)^n$
	$Hemicellulose \xrightarrow{r_2} \alpha_4 volatiles + \alpha_5 TAR + \alpha_6 char; \; r_2 = A_t exp\left(\frac{-E_i}{T_s}\right)(1-a_i)^n$
	$Lignin \xrightarrow{r_3} \alpha_7 volatiles + \alpha_8 TAR + \alpha_9 char; \; r_3 = A_t exp\left(\frac{-E_i}{T_s}\right)(1-a_i)^n$
	$Plastics \xrightarrow{r_4} \alpha_{10} volatiles + \alpha_{11} TAR + \alpha_{12} char; \; r_4 = \left[\sum_{i=1}^{n} A_i exp\left(\frac{-E_i}{RT}\right)\right]\rho_v$
	$PrimaryTAR \xrightarrow{r_5} volatiles + SecondaryTAR; \; r_5 = 9.55 \times 10^4 exp\left(\frac{-1.12\times10^4}{T_g}\right)\rho_{TAR1}$
Homogeneous reactions	$CO + 0.5O_2 \xrightarrow{r_6} CO_2; \; r_6 = 1.0 \times 10^{15} exp\left(\frac{-16000}{T}\right)C_{CO}C_{O_2}^{0.5}$
	$CO + H_2O \xleftrightarrow{r_7} CO_2 + H_2; \; r_7 = 2780 exp\left(\frac{-1510}{T}\right)\left[C_{CO}C_{H_2O} - \frac{C_{CO_2}C_{H_2}}{0.0265 exp\left(\frac{3968}{T}\right)}\right]$
	$CO + 3H_2 \xleftrightarrow{r_8} CH_4 + H_2O; \; r_8 = 3.0 \times 10^5 \, exp\left(\frac{-15042}{T}\right)C_{H_2O}C_{CH_4}$
	$H_2 + 0.5O_2 \xrightarrow{r_9} H_2O; \; r_9 = 5.159 \times 10^{15} \, exp\left(\frac{-3430}{T}\right)T^{-1.5}C_{O_2}C_{H_2}^{1.5}$
	$CH_4 + 2O_2 \xrightarrow{r_{10}} CO_2 + 2H_2O; \; r_{10} = 3.552 \times 10^{14} \, exp\left(\frac{-15700}{T}\right)T^{-1}C_{O_2}C_{CH_4}$
Heterogeneous reactions	$C + 0.5O_2 \xrightarrow{r_{11}} CO; \; r_{11} = 596 T_p exp\left(\frac{-1800}{T}\right)$
	$C + CO_2 \xrightarrow{r_{12}} 2CO; \; r_{12} = 2082.7 exp\left(\frac{-18036}{T}\right)$
	$C + H_2O \xrightarrow{r_{13}} CO + H_2; \; r_{13} = 63.3 exp\left(\frac{-14051}{T}\right)$

This model weights the effect of the Arrhenius rate and the diffusion rate of the oxidant at the surface particle. The diffusion rate coefficient is defined as follows:

$$D_0 = C_1 \frac{\left[(T_p + T_\infty) \div 2\right]^{0.75}}{d_p} \tag{2}$$

The final reaction rate is defined by:

$$\frac{dm_p}{dt} = -A_p \frac{\rho RT_\infty Z_{OX}}{M_{w,OX}} \frac{D_0 r_{Arrhenius}}{D_0 + r_{Arrhenius}} \tag{3}$$

4. Energy Analysis

In order to analyze the production of electricity from the selected substrates and to evaluate the efficiency of the process, an energy balance was developed to describe the gasification system in conjunction with the ICEG. Based on the principle of conservation of energy, the first law of thermodynamics, the energy balance can be written in a general form as follows:

$$\dot{Q} + \sum \dot{m}_{in}\left(h_{in} + \frac{V_{in}^2}{2} + gZ_{in}\right) = \dot{W} + \sum \dot{m}_{out}\left(h_{out} + \frac{V_{out}^2}{2} + gZ_{out}\right) \tag{4}$$

To simplify the analysis, a series of assumptions were followed. In the system described, the energy balance can be written as:

$$\dot{Q} + \sum \dot{m}_{in} \cdot h_{in} = \dot{W} + \sum \dot{m}_{out} \cdot h_{out} \tag{5}$$

where $\dot{Q}$ is the heat rate, $\dot{W}$ the work rate and h the specific enthalpy. At this point, the gasification system and the ICEG will be evaluated separately. For the gasification system, the energy balance is described in the following equations:

$$\dot{Q}_{Fuel} + \dot{Q}_{heat} = \dot{Q}_{syngas} + \dot{Q}_{tar} + \dot{Q}_{system} \tag{6}$$

The fuel's energy input can be calculated as:

$$\dot{Q}_{Fuel} = \dot{m}_{Fuel} \cdot LHV_{Fuel} \tag{7}$$

where LHV is the lower heating value and $\dot{m}_{Fuel}$ is the mass flow rate for fuel (biomass and MSW). As long as there is no condensation occurring the power of the supplied air can be given by:

$$\dot{Q}_{air} = \dot{m}_{air} \sum_{1}^{j} w_j Cp_j (T - T_0) \tag{8}$$

where w_j is the mass fraction and Cp_j the specific heat of a component, $(T - T_0)$ represents the preheat gas temperature and ambient temperature, respectively. To calculate the energy associated with the syngas ($\dot{Q}_{syngas}$) and tar ($\dot{Q}_{tar}$), the following equations are used:

$$\dot{Q}_{syngas} = \dot{m}_{syngas} \cdot LHV_{syngas} \tag{9}$$

$$\dot{Q}_{tar} = \dot{m}_{tar} \cdot LHV_{tar} \tag{10}$$

To calculate the energy efficiency of the cold gas, the useful output energy is divided between the energy from the input to the system, as follows:

$$\eta_{coldgas} = \frac{\dot{Q}_{syngas}}{\dot{Q}_{Fuel} + \dot{Q}_{air}} \tag{11}$$

The energy balance for the ICEG is configured as follows. The energy of the syngas that enters the ICEG is calculated by the Equation (12), shown below. The complete combustion of syngas depends on chemical composition. The general equation for the complete combustion of syngas with the theoretical amount of air required is given by:

$$C_X H_Y N_Z + a(O_2 + 3.76N_2) = bCO_2 + cH_2O + dO_2 + eN_2 \tag{12}$$

The values of the unknown coefficients a, b, c, d, and e, in the above equation, can be determined by applying the principle of conservation of mass to each element that constitutes the syngas [8].
The total heat loss in the ICEG is calculated by:

$$\dot{Q}_{lossICE} = \dot{Q}_{Syngas} - \dot{W}_{ICEG} \tag{13}$$

The thermal efficiency of the ICEG is generally determined as the ratio of the power output between the incoming fuel energy, as given by:

$$\eta_{ICE} = \frac{\dot{W}_{ICEG}}{\dot{Q}_{Syngas}} \tag{14}$$

The overall system efficiency is determined by:

$$\eta_{overall} = \frac{\dot{W}_{ICEG}}{\dot{Q}_{fuel}} \qquad (15)$$

5. Techno-Economic Analysis

Methodology

The economic analysis of biomass gasification is important and necessary to determine the cost of energy production and ensure the viability of the project. As an effort to bring this analysis closer to a real application scenario the present study was built based on literature review related with gasification investment projects [20,25]. The proposed 15 kW downdraft gasifying units coupled with an ICEG are set for deployment within a Portuguese and Brazilian decentralized community settled in a rural scheme. The small-scale units are set to be lodged within the community located near forest and agriculture landing for higher biomass availability with impact on transport costs and profitability of the supply operation. In Portugal, acacia is an invasive plant and estimates say that it occupies an area of approximately 43 thousand hectares, whereas in Brazil eucalyptus plantations reach around 7 million hectares [31,32]. An MSW supply chain is also considered arriving from surrounding urban areas and from the community itself. This supply chain profits from the already existing solid residues collection network to the urban areas and local communities. In Portugal, the average production rate of MSW is 1.32 kg per capita per day, making approximately 13.58 thousand tons of MSW per day, while in Brazil the average generation of MSW per capita is 1.15 kg of MSW per capita per day, approximately 234 thousand tons of MSW per day [20,33]. A near existing power line is assumed to connect the unit to the grid. The unit operation is assumed to be monitored by local workers already performing other tasks within the community, thus neither units require fully dedicated labor allowing to save on employees' costs [34]. For the economic evaluation, the main input financial data regulated in each country, Portugal and Brazil, such as input, capital and operating costs, alongside other key financial assumptions are provided in Table 6. For uniformity purposes, all costs were converted to euros (€).

Table 6. Main economic assumptions for the Portuguese and Brazilian proposed systems.

Item	Value	Remarks	Ref.
$I_{Gasifier}$ (€/kW)	2500	Costs related to the acquisition of the downdraft biomass gasifier, gas cleaning and conditioning system. Applied to both systems.	[27]
I_{ICEG} (€/kW)	500	Costs related to the acquisition of the internal combustion engine and electric generator. Applied to both systems.	[27]
$C_{Electricity\ Portugal}$ (€/kWh)	0.12	Electricity sales price practiced in Portugal.	[35]
$C_{Electricity\ Brazil}$ (€/kWh)	0.14	Electricity sales price practiced in Brazil.	[36]
$C_{O\&M}$ (%)	10	Operation and maintenance (O & M) costs refer to 10% of the total investment, applied accordingly to both systems. Includes salaries, electricity, water, ash removal and equipment maintenance costs.	[27]
C_{Acacia} (€/t)	35	Acacia acquisition costs for Portugal include transportation and conditioning (drying and splintering).	[25]
C_{MSW*} (€/t)	23	MSW acquisition costs for Portugal.	[37]
$C_{Eucalyptus}$ (€/t)	33	Eucalyptus acquisition costs for Brazil include transportation and conditioning (drying and splintering).	[31]
C_{MSW**} (€/t)	25	MSW acquisition costs for Brazil.	[20]
i (%)	12	Average discount rate considered per year. Applied to both systems.	[20]
Taxation (%)	15	Performance tax for small and medium-sized companies in mainland Portugal	[38]
	17	Performance tax for small and medium-sized companies in Brazil	[39]

The analysis of costs and income of cash flow before taxes (CFBT) considers, an initial investment period related to the phase of acquisition and assembly of the gasification system, as well as, the period of debt amortization and the costs related to the operation and maintenance (O & M). Revenues incur in electricity sales to the national grid. CFBT is calculated by balancing revenues and expenses while further applying the discount rate, as shown in Equation (16) [40]:

$$CFBT = \left(\sum Revenues - \sum Expenses \right) / (1+i)^t \tag{16}$$

Cash flows after taxes (CFAT), is one of the most useful liquidity measures to assess the financial health of a company since it considers the effect of the tax burden on the obtained profits. It also allows calculating the economic viability of the future investment while measuring the profitability or growth of an investment. CFAT is determined by the following equation, which relates CFBT minus taxation (Tax) [41]:

$$CFAT = CFBT - Tax \tag{17}$$

$$Tax = TXI{\cdot}TXR \tag{18}$$

$$TXI = CFBT - (DEP{\cdot}Inv.) \tag{19}$$

where, TXI represent the taxable income, TXR is the tax rate, Inv. is the initial investment, and DEP the depreciation, which is the amount that tax authorities allow to deduct from taxes. The depreciation of assets varies considerably from one country to another, in Portugal, a rate of 8.3% is considered for power generation companies [42], while in Brazil this rate is 10% [43].

In the case of Portugal, the calculation of taxes on profits and the fiscal incentives granted by the government for certain regions in the interior of the country were considered, coinciding with the area foreseen for the installation of this type of technology. Therefore, for the Portuguese scenario, a reduced tax rate of 15% was considered. As for Brazil, the tax burden is one of the highest in the world, over 32% [44], still, companies that generate energy from renewable sources (biomass included) benefit from a series of government tax incentives that reduce these charges up to 17% [39].

Cash flows after taxes along the project lifetime are applied to a spreadsheet-based economic model developed to calculate the net present value (NPV), internal rate of return (IRR) and the payback period (PBP).

The economic feasibility of the gasification system integrated with the ICEG is evaluated considering four distinct scenarios through the combination of three methods, NPV, IRR and PBP. According to Cardoso et al. [25], each of these methods carries its strengths and limitations. For example, NPV allows maintaining a good cash flow record, yet, a small increase or decrease in the discount rates significantly affects the results. On the contrary, IRR despite providing a simpler approach evaluates every investment by delivering merely one single discount rate. Regarding PBP, it presents an easily perceptible and direct analysis when calculating the amount of time required to recover from the investment, yet, it does not consider the financing and the risks associated with the venture. Beyond the weaknesses concerning each method, there is no doubt that each one fulfills a specific purpose in the economic analysis. Hence, in this study one provides an approach that allows evaluating the viability of the system by combining these three methods, further strengthening the economic model towards improved decision making. Additional details on the economic model formulation are provided in [25].

6. Results and Discussion

6.1. Experimental Runs and Model Validation

In order to validate the model a set of experimental data gathered from the 15 kWe downdraft gasifier was used for comparison. For validation purposes, as the research group already validated the here employed model in dealing with eucalyptus and MSW gasification in previously published works [8,45], and given the chemical similarity between Eucalyptus and MSW from both countries (Portugal and Brazil), it is then feasible to consider that the model proved to be sufficiently robust in dealing with these feedstocks. Therefore, in this work, only experimental gasification runs were performed for acacia residues to validate the model performance in dealing with this specific feedstock. In this sense, Table 7 shows the operating conditions for three experimental runs in which acacia residues were used as feedstock, while Figure 3 depicts the molar fraction as a function of gasification temperature for acacia.

Table 7. Experimental operating conditions and syngas analysis.

Experimental Conditions	Acacia Residues		
Run	1	2	3
Temperature (°C)	750	810	850
Biomass mass flow rate (kg/h)	15.8	15.8	11.5
Air flow rate (Nm^3/h)	18.0	18.0	28.0
Moisture (%)	14.2	14.2	14.2

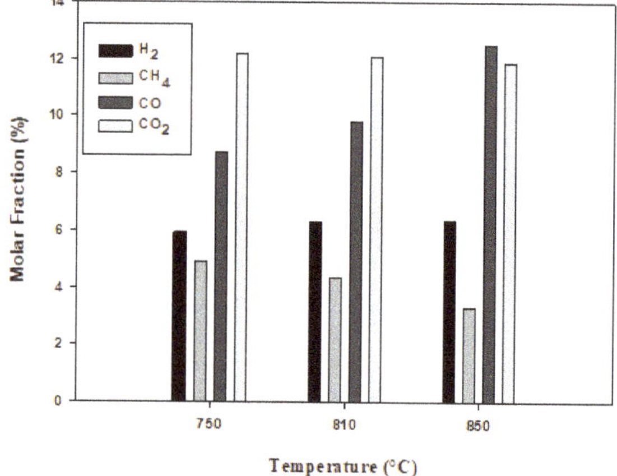

Figure 3. Molar fraction as a function of gasification temperature for the acacia residues.

Figure 4 compares the experimental and numerical syngas composition attained from the gasification runs of acacia residues. One can verify that the numerical results provide a good agreement with the results obtained from the experimental process carried out in the downdraft gasifier in dealing with the acacia residues. Considering all the simulations (for the four fuels), the maximum error obtained was inferior to 20%. This is a reasonable margin of error for such a complex system as biomass gasification. The differences between the model and the experimental results are due to some simplifying assumptions used in our model and to differences in the biomass fuel composition [46].

The numerical model successfully describes the syngas composition changes depending on the different operating conditions. In Figure 4a,b, the increase in the gasification temperature promote the formation of syngas with higher H_2 and CO content, as indicated by the experimental data. Both H_2 and CO appear to increase to asymptotic values. On the other hand, Figure 4c,d, show that the contents of CH_4 and CO_2 follow the opposite trend. CH_4 decreases as a function of temperature as the methane reaction is exothermic. These results are in close agreement with the literature [46]. According to Le Chatelier's principle, higher temperatures favor products in endothermic reactions. Therefore, the endothermic reaction in Table 5 (r13) was strengthened leading to an increase in the hydrogen content. A more detailed explanation of the model used in this work can be found elsewhere [30,46].

Finally, in this work, only three experimental runs were considered, once the numerical model applied has already been thoroughly validated concerning the syngas compositions attained from various reactors, multiple gasification agents and feedstocks at various operating conditions, strengthening the accurate predictability of the numerical model in a broad range of applications [28,47].

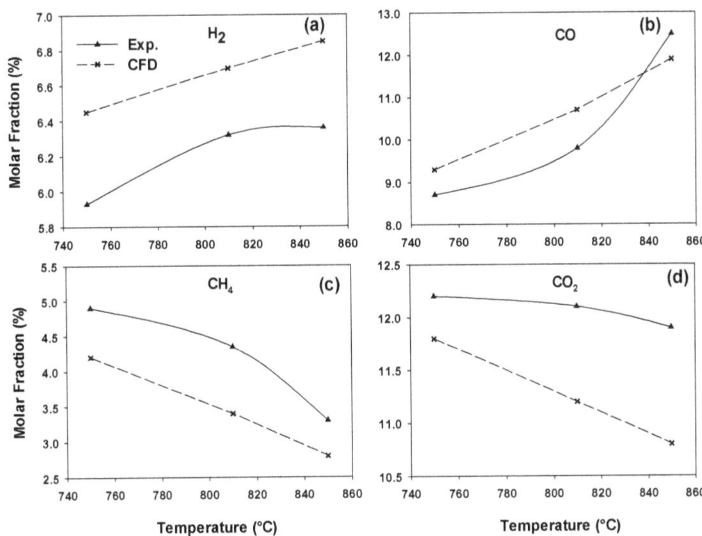

Figure 4. Experimental and numerical syngas composition deviations for acacia residues for the gas species: (**a**) H_2, (**b**) CO, (**c**) CH_4 and (**d**) CO_2.

6.2. Energy Analysis Results

To investigate the distribution of the energy flows of each stream in the gasification system-ICEG, the energy analysis was performed by the first law of thermodynamics. Table 8 shows the mass and energy balance for the analyzed feedstocks. The mass balance allows calculating the mass flow of the syngas and engine exhaust gases, as well as the air flows for the gasification process and syngas combustion. Biomass fuel consumption is assumed constant considering the nominal capacity of the gasification system, around 22 kg/h. Given the downdraft gasifier's hopper capacity, around 60 kg, the fuel residence time is about 3 h, depending on the used feedstock density and the air/ratio gas (equivalence ratio, ER).

Table 8. Main results of the technical analysis.

Acacia				MSW *			
Pts	$\dot{m}$ (kg/h)	P (kPa)	T (°C)	E (kW)	Pts	$\dot{m}$ (kg/h)	P (kPa)	T (°C)	E (kW)

Pts	$\dot{m}$ (kg/h)	P (kPa)	T (°C)	E (kW)	Pts	$\dot{m}$ (kg/h)	P (kPa)	T (°C)	E (kW)
1	22.0	101.3	25.0	103.9	1	22.0	101.3	25.0	88.0
2	59.4	101.3	25.0	-	2	48.1	101.3	25.0	-
3	81.4	101.3	25.0	59.1	3	70.1	101.3	25.0	64.2
4	0.01	101.3	25.0	-	4	0.03	101.3	25.0	-
5	440.1	101.3	25.0	-	5	428.3	101.3	25.0	-
6	-	-	-	10.64	6	-	-	-	11.6
7	521.48	101.3	450.0	63.1	7	493.0	101.3	450.0	52.6

Eucalyptus				MSW **			
Pts	$\dot{m}$ (kg/h)	P (kPa)	T (°C)	E (kW)	Pts	$\dot{m}$ (kg/h)	P (kPa)	T (°C)	E (kW)
---	---	---	---	---	---	---	---	---	---
1	21.8	101.3	25.0	111.5	1	20.4	101.3	25.0	111.1
2	41.2	101.3	25.0	-	2	64.8	101.3	25.0	-
3	63.0	101.3	25.0	83.3	3	85.2	101.3	25.0	83.3
4	0.02	101.3	25.0	-	4	0.03	101.3	25.0	-
5	483.7	101.3	25.0	-	5	435.4	101.3	25.0	-
6	-	-	-	15.0	6	-	-	-	15.0
7	546.7	101.3	450.0	68.3	7	520.6	101.3	450.0	68.3

The calculated values of the energy analysis for the downdraft-ICEG system given by the first law of thermodynamics are as follows. The power supplied by the biomass varies from 88 to 112 kW depending on the LHV of the fuel, the cooling efficiency of the gasifier varies from 73 to 75%, the power supplied by the syngas ranges from 64.2 to 83 kW. The electricity generated ranges between 11.6 and 15 kW, with an overall system efficiency of approximately 13.5%, as shown in Figure 5. It can be seen that the value of the overall efficiency of the system is in range with the values reported in the literature, approximately 14% [13,23].

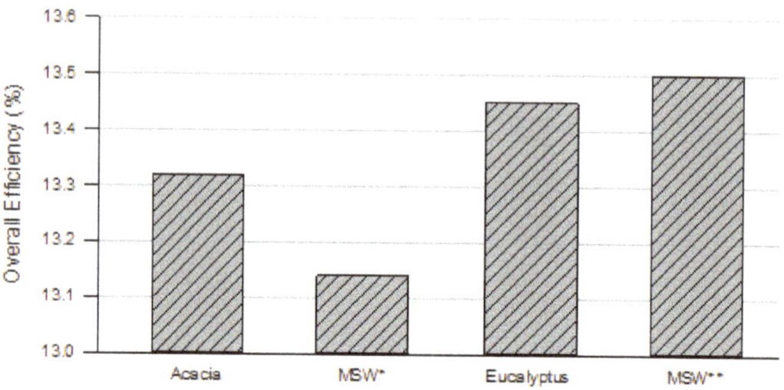

Figure 5. Overall efficiency for the four fuels considered.

Figure 6 shows the Sankey diagram of the energy flows for the proposed unit which includes the flow for the downdraft gasification system and the flow for the ICEG. The energy rates of each current vary accordingly to the LHV of the used fuel. The fuel processing capacity of the gasification system is limited to 22 kg/h, so fuels with lower LHV (acacia and MSW *) generate less energy. The energy contained in the fuel (biomass and MSW) is converted into physical and chemical energy of the syngas during the gasification process, thermal energy and solid waste become the main loss sources during this stage of the process, representing 26% for the acacia (Figure 6a), 27% for the MSW * (Figure 6b), 25.3% for the eucalyptus (Figure 6c) and 25% for MSW ** (Figure 6d). The syngas is combusted in an ICEG to generate electricity, the energy losses at this stage are the largest of the system, representing 60.7% for the acacia, 59.9% for the MSW *, 61.3% for the eucalyptus and 61.5% for MSW **. These losses are given mainly to the heat energy loss by the exhaust gases and the ICEG system. The overall efficiency of the system is approximately 14%. One possible route to reduce these losses is by implementing a cogeneration system that, in addition to producing electricity, will generate hot and cold water increasing the total efficiency of the system.

Figure 7 shows the fuel feeding rate effect in the overall efficiency of the downdraft-ICEG system for each one of the studied fuels. One can see that the overall efficiency ranges from 12.1 to 13.6%, setting the optimal values range for the fuel supply. The fuel feed value in the gasifier is approximately 22 kg/h, producing between 60 to 70 m³/h of syngas depending on the fuel.

The calculations made in this study show that it is technically and energetically possible to produce electrical energy through employing a downdraft gasifier in dealing with different fuels (biomass and MSW) so to produce syngas to feed an ICEG. However, syngas usage leads to engine power losses that can vary between 20 and 55% mainly in engines that were originally designed to operate with gasoline or natural gas [22]. These output engine power losses relate with the syngas LHV, the amount of combustible mixture supplied to the cylinder, and the number of combustion strokes in each time (number of revolutions per minute, rpm). For comparison purposes, Table 9 summarizes several studies also performed with downdraft gasifiers coupled with ICEG using 100% syngas.

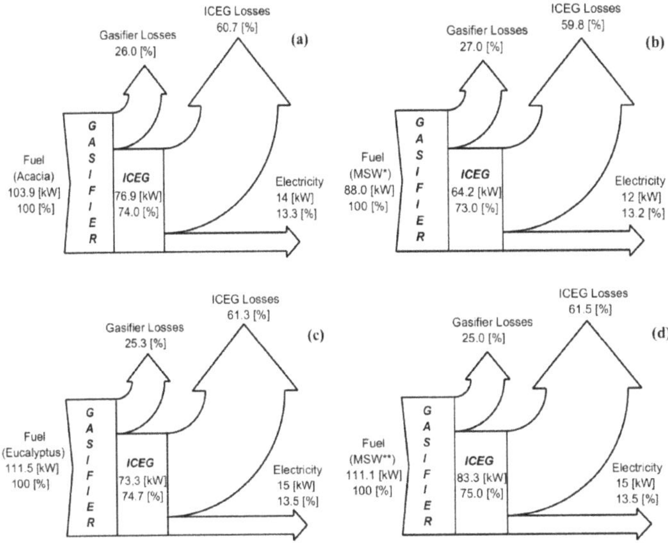

Figure 6. Sankey diagram of the gasification system-ICEG for: (a) acacia, (b) MSW *, (c) eucalyptus and (d) MSW **.

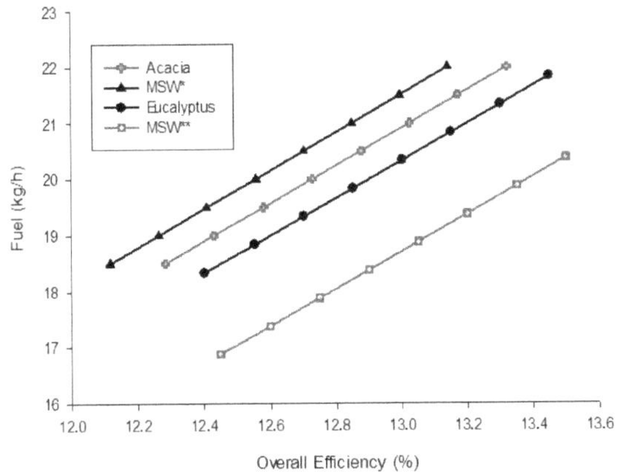

Figure 7. Effect of fuel feeding rate on the downdraft-ICEG system overall efficiency.

Downdraft-ICEG gasification systems can be applied to meet the energy needs of small communities or households, especially in difficult-to-reach rural settings. For instance, in Brazil, there are around 1 million residences without proper access to electricity, residences in which grid extension is unfeasible due to their limited access and low demand [4]. Thus, a viable route to electrify these residences is by implementing renewable energy solutions such as solar photovoltaics, small-hydroelectric, wind and small-scale gasification units. Among these solutions, downdraft-ICEG gasification systems are a feasible option for the Brazilian context as these residences are located in decentralized areas, of a rural or mountainous kind, where agriculture is a primary activity hence biomass availability is of no concern. Moreover, some of these communities dispose of easy access to MSW, either arriving from the community itself or even from surrounding urban areas, that when properly conditioned can be used to fuel the gasifying systems allowing the treatment and valorization of the organic fraction of this unsought product while maintaining a stable feedstock supply [13,25].

Table 9. Various studies performed with downdraft gasifiers coupled with ICEG using 100% syngas.

	Downdraft Gasifier Parameters				Engine Parameters							
Type of Biomass	Capacity (kg/h)	ER	Cold Gas Efficiency (%)	Syngas LHV (MJ/Nm^3)	Engine Size (kW)	Original Fuel Type	Original Engine Efficiency (%)	Performed Modifications	Maximum Brake Power Produced (kW)	Specific Fuel Consumption (kg/kWh)	Efficiency (%)	Ref.
Acacia, MSW *, Eucalyptus, MSW **	22	0.3	73 to 75	3 to 4.5	20	Gasoline; GN	28 to 30	Adapted carburetor	13 to 15	3 to 4	13.5 (Overall efficiency)	This study
Eucalyptus	20 to 30	N/A	69	5.5	10	Gasoline; NG	28 to 32	Adapted carburetor	5.2	N/A	12.8 (Electrical efficiency Gasifier -ICEG)	[13]
MSW	140	0.3	75	4.6	96	Gasoline	33	N/A	97	3.5	23 (Electrical efficiency ICEG)	[20]
Wood chips	87	0.3 to 5	88	5.6	100	NG	N/A	Adapted carburetor	73	3.21	21 (Overall efficiency)	[21]
Switchgrass	100	0.2 to 5	80	6 to 7	10	NG	28	Air fuel intake	5	1.9	21.3 (Electrical efficiency ICEG)	[22]
Longan tree charcoal	5 to 6	N/A	N/A	4.64	8.2	Diesel	30 to 35	Converted to SI engine, develop new carburetor	3.17	5.53	23.5 (Mechanical efficiency ICEG)	[48]

N/A: Not available; NG: Natural Gas.

Regarding Portugal, most of the forest biomass availability is located mainly in the country's central region, out of which nearly 97.5% of the woodlands belong to private owners or local communities [33]. In 2017, deadly wildfires flared across central and north of Portugal, calling for an articulated set of actions promoting active forest management focusing on increasing the value of forest products through the use of biomass for energy purposes [25].

Thus, in this follow-up, the implementation of small power generation systems, such as the one analyzed in this study, suits as a viable option for the valorization of local forest biomass residues for decentralized communities, while preventing wildfire hazard and maintaining a biomass-to-energy strategy and policy measures impacting forest products trade. Again, acknowledging the gasification with local MSW stands as a clever strategy to increase the unit's production efficiency aiding towards achieving the energy independence of these remote communities.

6.3. Economic Model Results

The economic analysis provides the financial viability of an investment project throughout its lifetime. Figure 8 illustrates the initial outlay for the 15 kWe downdraft gasifier-ICEG project, set around 45,000 € for a lifetime of 25 years, jointly with the calculated cash flows, total annual outflows and inflows. During the investment period (year 0), the expenses required to initiate the project are thought-through. For year 1, the project's cash flow increases to positive status due to electricity sales income as the plant starts generating electric power. It is assumed that starting this year the plant will generate approximately 36,000 kWh/year for the rest of the project's lifetime, achieving an average profit rate of 10,500 €/year. In addition to the initial investment (only reflected in year 0), outflows result from expenses related to fuel purchase (biomass or MSW) and O&M costs. Expenses related to fuel represent approximately 85% of the total expenses of the plant. Revenues show a positive flow over the years, emphasizing the importance of the electricity sales price and reimbursements to the projects' viability. The project installation will be completely debt-free in 9.20 years for the acacia option (as seen in Figure 8a), 12.61 years for the MSW * (Figure 8b), 9.38 years for the eucalyptus option (Figure 8c) and 8.67 years for MSW ** (Figure 8d). These assumptions go hand in hand with the PBP calculations, as explained below. The present project cash flows distribution results are in accordance with the assumptions and recommendations found in the literature [25,49]

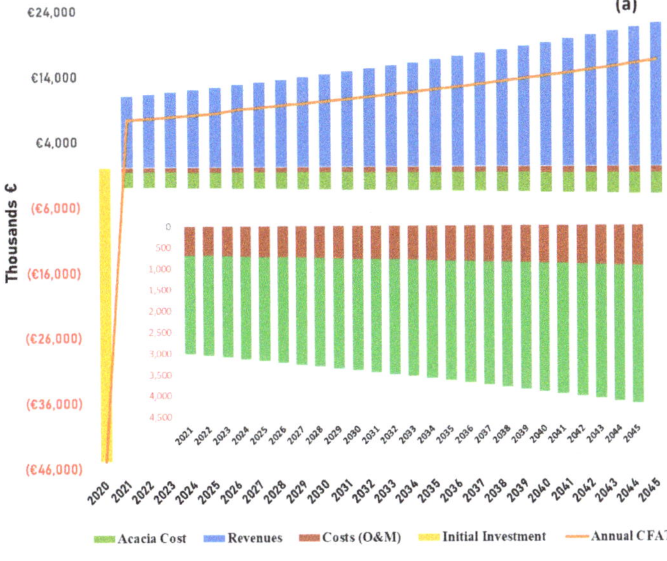

Figure 8. *Cont.*

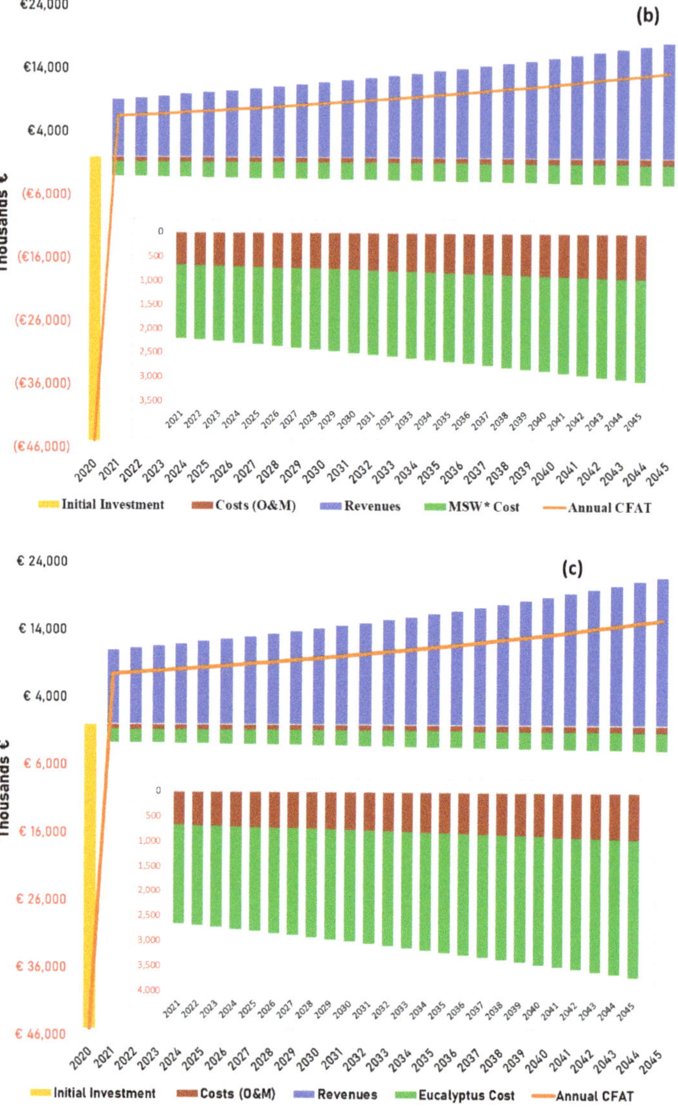

Figure 8. *Cont.*

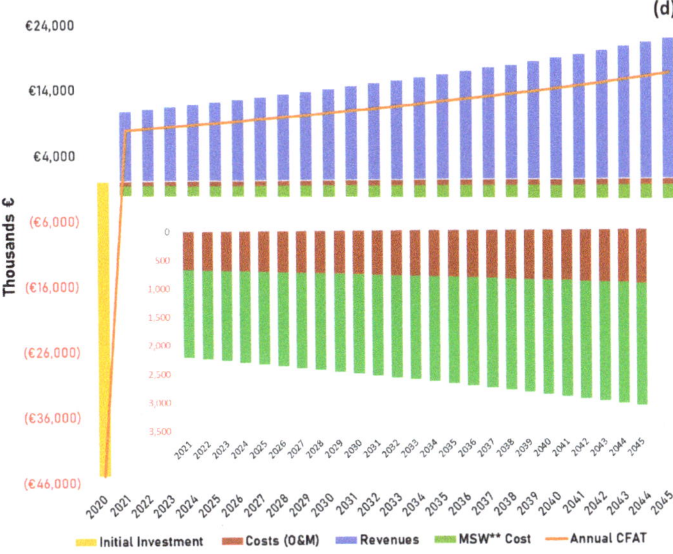

Figure 8. Cash flows concerning the various gasification projects depicting the annualized costs and revenues throughout the plant's lifetime for Portugal with (**a**) acacia and (**b**) MSW *, and Brazil with (**c**) eucalyptus and (**d**) MSW **.

After calculating the cash flows, one must determine if the project is economically acceptable. Figure 9 provides the economic evaluation results for the NPV, IRR and PBP calculations. The NPV is calculated concerning a discount rate of 12%, which will determine whether the project is to be accepted or not. For higher discount rates the NPV value decreases, while for lower discount rates the NPV value increases, showing that the NPV behavior is inversely proportional to the discount rate. Results show that the NPV for the downdraft-ICEG system project dealing with acacia is of 29.32 k€ (Figure 9a), 18.99 k€ for MSW * (Figure 9b), 28.45 k€ for eucalyptus (Figure 9c) and 31.65 k€ for MSW ** (Figure 9d). The IRR rate is 19.34% for acacia, 16.88% for MSW *, 19.28% for eucalyptus and 20.09% for MSW **. Please note that the IRR is calculated concerning the CFAT results and not on the CFBT results. Lastly, the PBP varies from 8.67 to 12.61 years according to the fuels considered. A closer look at the results shows that the investment projects set for deployment in Brazil deliver enhanced economic performances showing increased feasibility. Such behavior is due mainly to the higher electricity sales price practiced in Brazil as the projects' revenues arrive from electricity sales to the national grid.

The financial indicator results obtained for the energy generation system projects are consistent with the current literature [13,20]. According to the World Bank Group, typical benchmarks for key financial parameters in biomass projects indicate that the NPV should be a positive value, the IRR should be higher than 10% and the PBP less than 10 years [49]. Bearing in mind these considerations, the projects here purposed may indeed be considered economically feasible. However, despite its apparent viability, one must go beyond these numbers and assess how economically attractive the venture may be to most investors and which are the main risks associated with it. In this sense, a sensitivity analysis must be performed so to determine which variables are the most critical to the project's success.

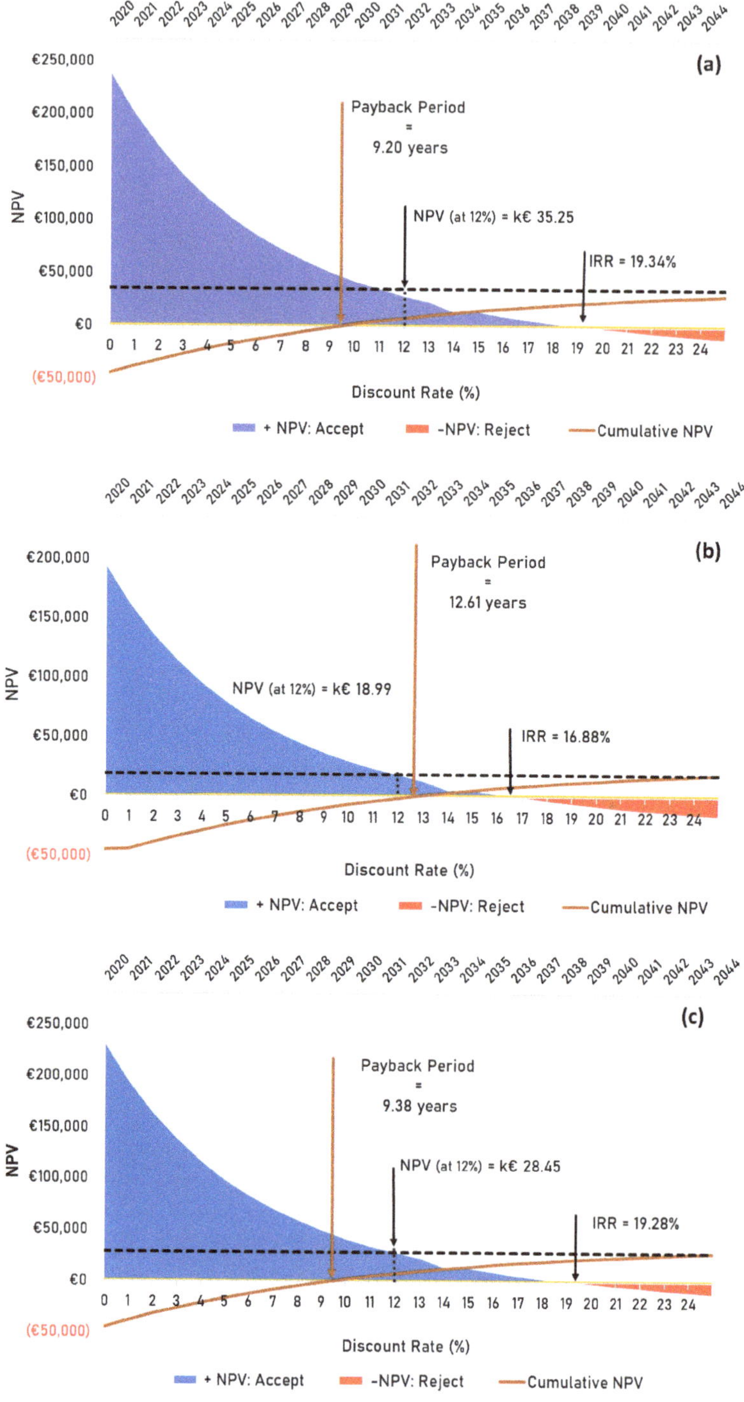

Figure 9. *Cont.*

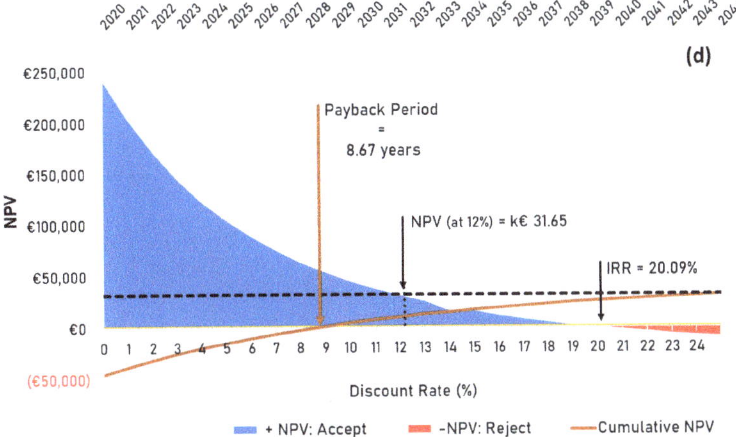

Figure 9. Financial indicators overview throughout the gasification power plant lifetime for (**a**) acacia, (**b**) MSW *, (**c**) eucalyptus and (**d**) MSW **.

7. Sensitivity Analysis

7.1. Monte Carlo Simulation

The sensitivity analysis evaluates the risk impact by varying each parameter while all other variables remain constant. This approach will enable changes in the results of the economic analysis, indicating the importance of specific risk parameters while assessing its influence on the system's performance [49]. The risk analysis was evaluated by means of the Monte Carlo method implemented within the economic model spreadsheet. A Monte Carlo method was employed for a total of 10,000 iterations following a triangular distribution considering a lower, upper and reference value. The sensitivity analysis moves in a range of ±30% over the reference value of the selected input variables. The input variables considered as the most influential in the project results are the initial investment, discount rate, electricity sales price, O & M and fuel costs. The input variables chosen coincide with those indicated as the most critical in the literature [25,49].

7.2. Sensitivity Analysis Results and Discussion

Figure 10a–d show the values for the NPV probability distribution for the four analyzed fuels respecting each system settled in both Portugal and Brazil. The mean distribution values, given by the 'Mean' vertical line in the NPV probability distributions figures, are very close to the average values previously calculated in the economic analysis. This closeness supports the validity of the simulations. The low standard deviation (referred to as St. dev.) indicates that the expected values within the distribution are close to the mean. In this sense, the higher the standard deviation is, the greater the investment risk will be. The highest standard deviation values are presented for acacia σ = 15.62 k€ (Figure 10a), eucalyptus σ = 15.18 k€ (Figure 10c) and MSW ** σ = 15.07 k€ (Figure 10d), and the lowest value for MSW * σ = 14.48 k€ (Figure 10b). Therefore, a higher probability of investment risk loss in the NPV is more likely to occur for the acacia, followed by eucalyptus and MSW **, and finally, MSW * which gathers the lowest risk of investment loss. Describing the NPV projection, the probability of attaining a negative NPV is given by the probability density area at the left of zero, for acacia and MSW * these are placed around 1.82% and 7.52%, respectively, while for eucalyptus and MSW ** these are set around 1.53% and 0.66%. This means that there is a rather low probability of reaching a negative NPV. Whereas the probability of this investment exceeding an NPV of 50 k€ is approximately 10.56% for acacia, 2.90% for MSW *, 9.65% for eucalyptus and 12.59% for MSW **.

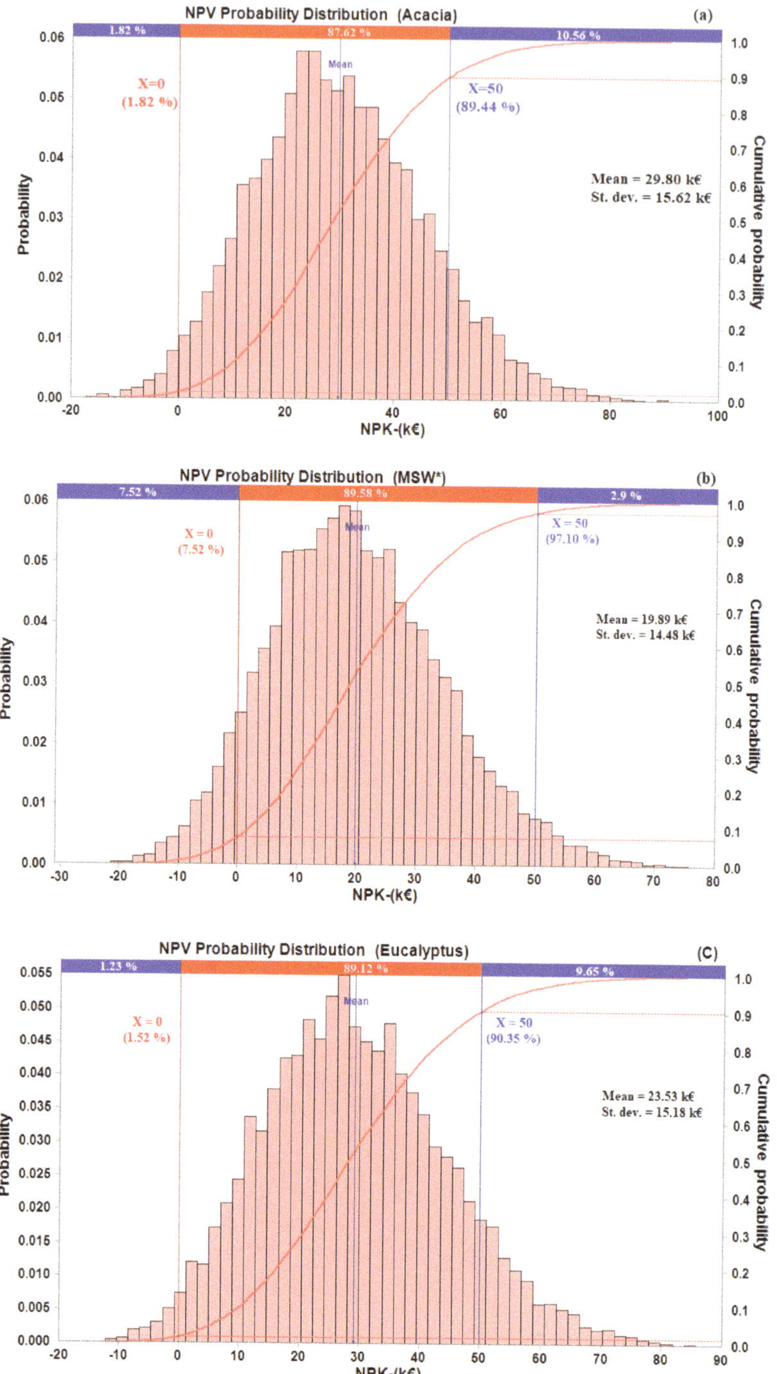

Figure 10. *Cont.*

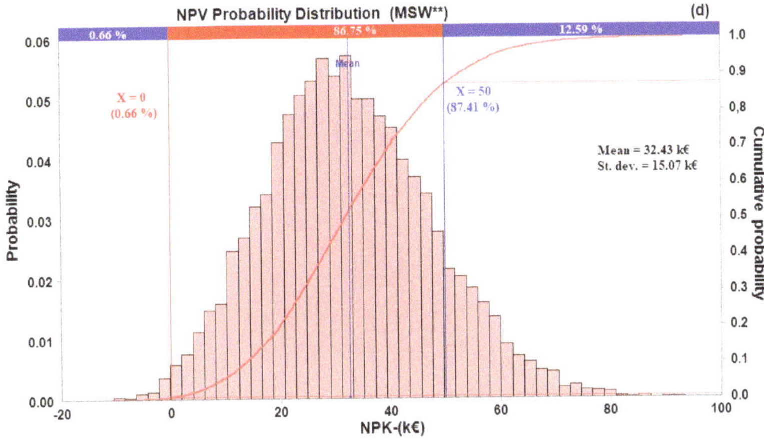

Figure 10. Probability distributions for NPV: (**a**) Acacia, (**b**) MSW *, (**c**) Eucalyptus and (**d**) MSW **.

In general, the NPV showed wider distribution over its mean value resulting in the highest standard deviation values as compared to IRR and PBP. Higher standard deviation values mean that higher investment loss is more likely to occur in the NPV. Therefore, for the sake of simplicity, instead of providing individual figures for all the studied fuels for both IRR and PBP, Table 10 is provided summarizing the mean and standard deviations values for the IRR and PBP probability distributions. In contrast to the NPV, both IRR and PBP carry lower standard deviation values of about 0.03% for IRR and between 3.21 to 5.52% for PBP, indicating that all their expected values tend to be very close to the mean, signifying that the lowest investment risk failure comes associated with these financial indicators, especially with the IRR, delivering the lowest values.

Table 10. IRR and PBP probability distributions for the four studied fuels.

Fuel	IRR		PBP	
	Mean (%)	St. dev (%)	Mean (Years)	St. dev. (Years)
Acacia	19.31	0.03	10.35	3.90
MSW *	17.05	0.03	13.95	5.52
Eucalyptus	19.46	0.03	10.23	5.30
MSW **	20.32	0.03	9.37	3.21

From all the financial indicators studied, NPV is the highest risk facing indicator as it presents the highest standard deviations. Hence, one will focus on discussing how the uncertainty of the input variables considered affects the NPV for each fuel. Figure 11 shows the impact of each input variable on the NPV and its sensitivity range. Results show that the impact changes on NPV are identical in the systems settled in Portugal and Brazil, as the profitability of the projects is highly sensitive to the electricity sales price and discount rate for all analyzed fuels. The next most striking input variables are mainly the initial investment, followed by O & M costs and fuel costs (as generally the least affecting). The electricity sale price and discount rate show a considerable impact change on the NPV as compared to the remaining input variables.

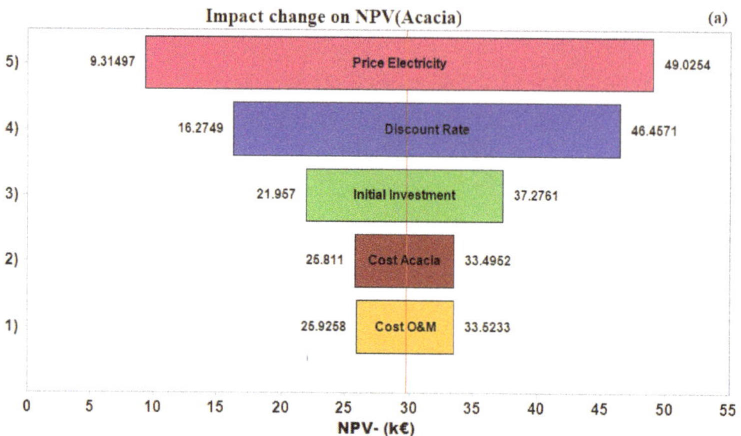

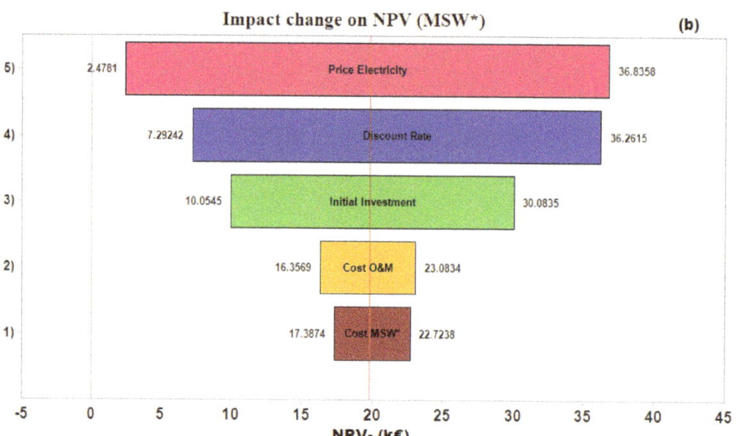

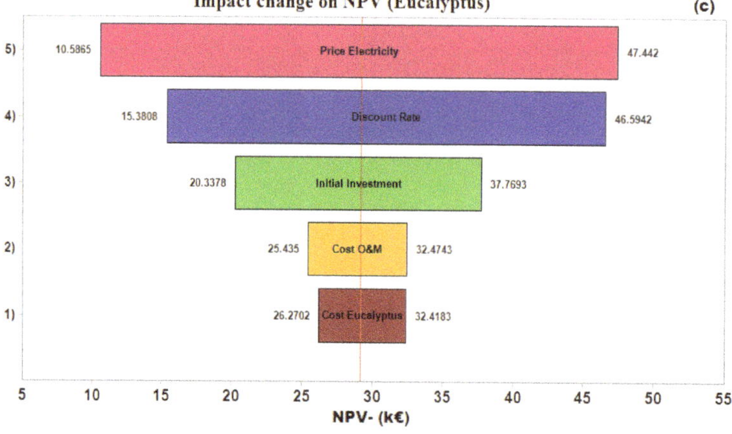

Figure 11. *Cont.*

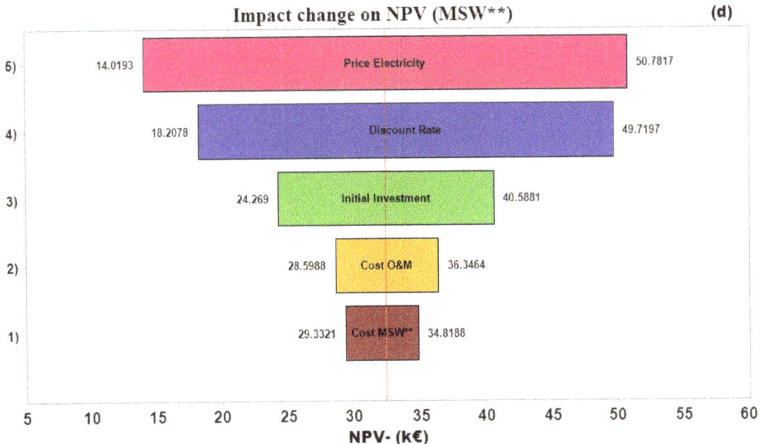

Figure 11. Sensitivity range to input variables for NPV for the different fuels: (**a**) acacia, (**b**) MSW *, (**c**) eucalyptus and (**d**) MSW **.

Out of all variables considered, the electricity sales price can considerably hamper the economic viability of the project, when considering an unfavorable scenario, the NPV can reach low values for the MSW * 2.48 k€ (Figure 11b). On the other hand, it can also considerably increase the income of the project into the most favorable scenario with acacia attaining 49.03 k€, MSW * 36.84 k€, eucalyptus 47.44 k€, and MSW ** 50.78 k€. The electricity sales price is a variable of extreme importance significantly affecting the viability of the projects as it is a rather uncertain parameter due to the energy market price fluctuations and subsidies, both highly dependent upon political decisions. The discount rate is the second variable with the greatest impact on the NPV, such is true given to the high impact in the NPV discounted cash flow calculations. As for the initial investment, it can be an impactful variable since it greatly influences the calculation of the NPV given the high initial cash flow of the project. Following, the O & M variable carries slither impact risk. This variable heights-in all costs necessary to maintain the system fully operational. Finally, and even in the most worsening scenario, the fuel cost is the variable that usually least affects the NPV. Ultimately, despite not being directly accounted for in the sensitivity analysis, in a real scenario, and according to the energy results, the increased LHV attributed to the Brazilian feedstocks would assist both the economic feasibility and risk assessment, in the long run, by allowing increased system performance.

Unsurprisingly, and even despite the narrow difference, the results of the economic analysis show that the costs of energy generation based on biomass gasification systems do vary accordingly with the country's economic scenario, as regulated expenses, revenues and fuel composition vary widely from one country to the other. To compare the results here attained with the ones found in the literature, Table 11 provides a record of economic analysis results from several previous works set in different countries employing power generation systems similar to the one set in this study. Indeed, the economic analysis results are in accordance with a range of values present in the literature under similar circumstances.

Table 11. Record of downdraft gasifiers' economic analysis results found in the literature suiting as means to of comparison for this work.

Fuel	Power (kW)	Discount Rate (%)	NPV (k€)	IRR (%)	PBP (Years)	Ref.
Acacia	13.8		35.25	19.34	9.20	
MSW *	11.6		18.89	16.88	12.61	Present
Eucalyptus	15.0	12	28.45	19.28	9.38	study
MSW **	15.0		31.65	20.09	8.67	
Olive tree pruning	70	N/A	302	N/A	5	[50]
Eucalyptus	10	12	N/A	N/A	5	[13]
Pellet	48	7	81	10.2	8	[16]
Agricultural Residues	1000	N/A	3250	18.1	7.8	[51]

N/A: Not available.

8. Considerations on Small-Scale Biomass Gasification vs. Conventional Diesel Solutions

Contextually, small-scale biomass gasification-ICEG systems granted a good fit for both Portuguese and Brazilian energy policies when applied to electric power generation in decentralized areas, particularly when settled in small municipalities of rural kind aiding to reduce the energy dependence from external sources.

So far, small-scale decentralized electrification solutions addressed to decentralized communities have been dominated by conventional diesel generators [52]. The world's growing demand for cleaner energy calls however for an urgent and sustainable low-carbon expansion in the power sector. As known, fossil fuel-driven engines emit NO_x, CO, unburned hydrocarbons and particulate matter, which contribute to air pollution and are particularly harmful to local inhabitants. Moreover, electric power production from these systems may fare expensive due to high diesel fuel costs, hampered by fuel price fluctuations and additional costs related to fuel transportation to remote sites with sparsely developed road structure. Plus, fuel transportation must be effectively planned to the point of safeguarding fuel supply, avoiding shortages or interruptions, which, in the long run, will lead communities into facing increased energy costs [53].

Renewable energy sources solutions such as biomass gasification systems coupled with ICEG, are proving to be a far more clean, reliable and efficient option for decentralized electrification, not requiring regular fuel supply and being cost-competitive with diesel generators [54]. In fact, a well operated and established biomass gasification system can produce far less greenhouse gas emissions since biomass is considered carbon-neutral, feedstocks arrive from renewable sources and most of the produced gas is used as fuel [55]. Therefore, the biomass systems' environmental performance and technical competitiveness with conventional diesel systems alongside with the current global warming paradigm upholds gasification technology to a frontline position shift towards a more sustainable energy market.

9. Conclusions

This work presented a comparative techno-economic analysis concerning the gasification of four fuels from two distinct countries, acacia and MSW * from Portugal and eucalyptus and MSW ** from Brazil, employed for electricity generation purposes. Experimental gasification runs for acacia were carried out in a 15 kWe gasifier manufactured by All Power Labs coupled with an ICEG. The syngas composition was predicted by employing a 2-D Eulerian-Eulerian approach developed within the ANSYS Fluent framework. The mathematical model was able to predict the experimental syngas composition within a reasonable 20% error, considering the complexity associated with biomass and MSW gasification processes.

The technical evaluation focused mainly on electricity generation and system efficiencies. Main energy analysis results settled the fuel power supply between 88 to 112 kW, depending on the LHV, the gasifier's cooling efficiency varied from 73 to 75% and the syngas power supply from 64.2 to 83 kW.

Ultimately, the generated electricity ranged from 11.6 to 15 kW with an overall system efficiency of approximately 13.5%. Hereupon, calculations showed the adequacy of producing electric power by means of a downdraft gasifier dealing with different fuels to produce proper syngas to feed an ICEG.

Regarding the economic analysis, a spreadsheet-based economic model was developed based on a combination of three financial indicators, NPV, IRR and PBP. The main cost factors considered included electricity generation costs, initial investment, O & M cost and fuel costs. Revenues were calculated from the electricity sales to the national grid. The economic model presented positive perspectives admitting the possibility of establishing the project under current market conditions. For the Portuguese scenario, the results showed an NPV of 29.32 k€ for the acacia and 18.99 k€ for MSW *, whereas, for the Brazilian scenario, eucalyptus resulted in an NPV of 28.45 k€ and 31.65 k€ for MSW **. Concerning the IRR, results showed rates of 19.34% for acacia, 16.88% for MSW *, 19.28% for eucalyptus and 20.09% for MSW **. As for PBP, results predicted a time frame of 9.20 years for acacia, 12.61 years for MSW *, 9.38 years for eucalyptus and 8.67 years for MSW **. In general, the investment projects set for deployment in Brazil delivered enhanced economic performances showing increased feasibility mainly due to higher electricity sales price practiced in the country.

A sensitivity analysis was carried out by employing the Monte Carlo method to assess the risks and measure the level of uncertainty associated with the ventures. The risk assessment yielded rather favorable investment projections with a probability of the NPV reaching positive values of 98.18% for acacia, 92.48% for MSW *, 98.77% for eucalyptus and 99.34% for MSW **. As compared to the NPV, IRR and PBP carried lower standard deviation values, of about 0.03% and 3.21 to 5.52%, respectively, signifying that the lowest investment risk failure comes associated with these two financial indicators, especially with the IRR. Hence, a greater risk of investment loss was detected for the NPV. Overall, the NPV showed to be considerably more sensitive to uncertainties associated with the electricity sales price and discount rate, while fuel cost was generally the least affecting variable. Finally, the least affected NPV given out by the Brazilian feedstocks showed that the units set for deployment in this context offer a slight lower risk of investment loss as compared to the Portuguese projects mostly due to improved electricity sales price regulations.

At last, this work states the positive effect of energy generation from small-scale gasification-ICEG systems in different scenarios (Portugal and Brazil) particularly towards decentralized communities inserted in a rural scheme. However, despite the broad benefits shown by this technology, its viability is still highly influenced by local government decisions, who must enforce light-handed regulatory measures concerning subsidiary support and favorable electricity sales prices so to benefit the dissemination and development of this technology. In the long run, these policies will not only allow investors to face the high initial outlays required for deploying such ventures, but also stimulate research interest in further developing gasification technology coupled with ICEG to a wide commercial viability status. Truly, small-scale gasification systems for decentralized solutions provide the opportunity for achieving global access to electricity, suiting as key for unlocking a sustainable future while uplifting the local economy in these locations.

Author Contributions: Conceptualization, V.S. and J.L.S.; methodology, C.E.T. and P.B.; software, D.E., J.C. and J.R.C.; validation, D.E., J.C. and J.R.C.; formal analysis, C.E.T., R.A.M.B. and V.S.; investigation, J.R.C.; resources, P.B. and J.L.S.; data curation, R.A.M.B. and C.E.T.; writing—J.R.C. and J.C.; writing—review and editing, J.R.C.; J.C., V.S. visualization, J.R.C.; supervision, R.A.M.B., J.L.S. and V.S.; project administration, V.S. and J.L.S.; funding acquisition, V.S. and J.L.S. All authors have read and agreed to the published version of the manuscript.

Funding: This research was funded by Coordination for the Improvement of Higher Education Personnel (CAPES) and by the Portuguese Foundation for Science and Technology (FCT) through the project number 88881.156267/2017-01, entitled Uso de Misturas Syngas/Biodiesel em MCI para a Geração Descentralizada de Energia. This paper is also a result of the project "Apoio à Contratação de Recursos Humanos Altamente Qualificados" (Norte -06-3559-FSE-000045), supported by Norte Portugal Regional Operational Programme (NORTE 2020), under the PORTUGAL 2020 Partnership Agreement.

Acknowledgments: The authors would also like to thank to the FCT for the grant SFRH/BD/146155/2019 and for the project IF/01772/2014.

Conflicts of Interest: The authors declare no conflict of interest. The funders had no role in the design of the study; in the collection, analyses, or interpretation of data; in the writing of the manuscript, or in the decision to publish the results.

Nomenclature

A, B	calibration constants
B_i	equation coefficient related to the main factors
B_0	interception coefficient
$B_{i,i}$	quadratic effects (give the curvature to the response surface)
$B_{i,j}$	cross interactions between factors
$C_{1\varepsilon}, C_{2\varepsilon}, C_{3\varepsilon}$	constants
C_p	specific heat capacity
Ct	cash flows
D_0	diffusion rate coefficient
G_k	generation of turbulence kinetic energy due to the mean velocity gradients
G_b	generation of turbulence kinetic energy due to buoyancy
g	gravitational acceleration
h	specific enthalpy
h_{pq}	heat transfer coefficient between the fluid phase and the solid phase
I	radiation intensity
J_i	diffusion flux of specie i
k	thermal conductivity
LHV	lower heating value
M	total mole flow of carbon in the syngas components
M_c	molecular weight
$M_{w,i}$	molecular weight of i component
m	biomass flow into the gasifier
$\dot{m}$	mass flow
$\dot{m}_{pq}$	mass flow between the fluid phase and the solid phase
Nu	Nusselt number
P_s	particle phase pressure due to particle collisions
p	gas pressure
$\vec{q}_q$	heat flux
qth	specific enthalpy
Q_{pq}	heat transfer intensity between phases
$\dot{Q}$	heat rate
R	universal gas constant
R_i	net generation rate of specie i due to homogeneous reaction
R_c	reaction rate
S_i	source term related to the specie i production from the solid heterogeneous reaction
S_k	user-defined source terms
S_q	source term due to chemical reactions
S_ε	user-defined source terms
T	temperature
U	mean velocity
v	instantaneous velocity
$\dot{W}$	work rate
X_c	carbon fraction in the biomass (obtained from the ultimate analysis)
Y	mass fraction
Y_m	contribution of the fluctuating dilatation in compressible turbulence to the overall dissipation rate

Other symbols

α	volume fraction
β	gas-solid interphase drag coefficient
ρ	density
Θ_s	granular temperature, proportional to the kinetic energy of the random motion of the particles
φ_{ls}	energy exchange between the fluid phase and the solid phase
$k_{\Theta_a}\nabla\Theta_s$	diffusion energy (k_{Θ_a} is the diffusion coefficient)
$\left(-P_s\bar{I}+\bar{\tau}_s\right):\nabla\left(\vec{v}_s\right)$	generation of energy by the solid stress tensor
γ_{Θ_a}	collisional dissipation of energy
τ	tensor stress
μ	viscosity
γ_c	stoichiometric coefficient
η	efficiency

Subscripts

g	gas phase
s	solid phase
i	component

References

1. Alsharif, M.H. Techno-economic evaluation of a stand-alone power system based on solar/battery for a base station of global system mobile communication. *Energies* **2017**, *10*, 392. [CrossRef]
2. Pardo, J.E.; Mejías, A.; Sartal, A. Assessing the importance of biomass—Based heating systems for more sustainable buildings: A case study based in Spain. *Energies* **2020**, *13*, 1025. [CrossRef]
3. APREN. *Eletricidade Renovável. Panorama Energético Nacional*; APREN: Lisboa, Portugal, 2017.
4. Ministry of Mines and Energy. *Resenha Energética Brasileira*; Ministry of Mines and Energy: Brasilia, Brazil, 2018.
5. Dudley, B.; Dale, S. BP Energy Economics. In *BP Energy Outlook*; BP: London, UK, 2018.
6. Uddin, M.N.; Techato, K.; Taweekun, J.; Rahman, M.M.; Rasul, M.G.; Mahlia, T.M.I.; Ashrafur, S.M. An overview of recent developments in biomass pyrolysis technologies. *Energies* **2018**, *11*, 3115. [CrossRef]
7. Ferreira, S.; Monteiro, E.; Calado, L.; Silva, V.; Brito, P.; Vilarinho, C. Experimental and modeling analysis of brewers' spent grains gasification in a downdraft reactor. *Energies* **2019**, *12*, 4413. [CrossRef]
8. Couto, N.; Silva, V.; Monteiro, E.; Teixeira, S.; Chacartegui, R.; Bouziane, K.; Brito, P.S.D.; Rouboa, A. Numerical and experimental analysis of municipal solid wastes gasification process. *Appl. Therm. Eng.* **2015**, *78*, 185–195. [CrossRef]
9. Rinaldini, C.A.; Allesina, G.; Pedrazzi, S.; Mattarelli, E.; Savioli, T.; Morselli, N.; Puglia, M.; Tartarini, P. Experimental investigation on a Common Rail Diesel engine partially fuelled by syngas. *Energy Convers. Manag.* **2017**, *138*, 526–537. [CrossRef]
10. Thi, T.; Huong, T.; Khanh, N.D.; Luong, P.H.; Tuan, L.A. A computational study of the effects of injection strategies on performance and emissions of a Syngas/Diesel dual-fuel engine. In Proceedings of the 5th AUN/SEED-Net Regional Conference on Global Environment, Bandung, Indonesia, 21–22 November 2012; pp. 379–392.
11. Guo, H.; Neill, W.S.; Liko, B. The combustion and emissions performance of a syngas-diesel dual fuel compression ignition engine. In Proceedings of the ASME 2016 Internal Combustion Engine Fall Technical Conference, Greenville, SC, USA, 9–12 October 2016. ICEF 2016.
12. Omar, M.M.; Munir, A.; Ahmad, M.; Tanveer, A. Downdraft gasifier structure and process improvement for high quality and quantity producer gas production. *J. Energy Inst.* **2018**, *91*, 1034–1044. [CrossRef]
13. Boloy, R.A.M.; Silveira, J.L.; Tuna, C.E.; Coronado, C.R.; Antunes, J.S. Ecological impacts from syngas burning in internal combustion engine: Technical and economic aspects. *Renew. Sustain. Energy Rev.* **2011**, *15*, 5194–5201. [CrossRef]
14. Dasappa, S.; Subbukrishna, D.N.; Suresh, K.C.; Paul, P.J.; Prabhu, G.S. Operational experience on a grid connected 100 kWe biomass gasification power plant in Karnataka, India. *Energy Sustain. Dev.* **2011**, *15*, 231–239. [CrossRef]

15. Lee, U.; Balu, E.; Chung, J.N. An experimental evaluation of an integrated biomass gasification and power generation system for distributed power applications. *Appl. Energy* **2013**, *101*, 699–708. [CrossRef]
16. Elsner, W.; Wysocki, M.; Niegodajew, P.; Borecki, R. Experimental and economic study of small-scale CHP installation equipped with downdraft gasifier and internal combustion engine. *Appl. Energy* **2017**, *202*, 213–227. [CrossRef]
17. Cardoso, J.; Silva, V.; Eusébio, D.; Brito, P. Hydrodynamic modelling of municipal solid waste residues in a pilot scale fluidized bed reactor. *Energies* **2017**, *10*, 1773. [CrossRef]
18. Silva, V.; Rouboa, A. Combining a 2-D multiphase CFD model with a Response Surface Methodology to optimize the gasification of Portuguese biomasses. *Energy Convers. Manag.* **2015**, *99*, 28–40. [CrossRef]
19. Coronado, C.R.R.; Yoshioka, J.T.T.; Silveira, J.L. Electricity, hot water and cold water production from biomass. Energetic and economical analysis of the compact system of cogeneration run with woodgas from a small downdraft gasifier. *Renew. Energy* **2011**, *36*, 1861–1868. [CrossRef]
20. Luz, F.C.; Rocha, M.H.; Lora, E.E.S.; Venturini, O.J.; Andrade, R.V.; Leme, M.M.V.; Del Olmo, O.A. Techno-economic analysis of municipal solid waste gasification for electricity generation in Brazil. *Energy Convers. Manag.* **2015**, *103*, 321–337. [CrossRef]
21. Raman, P.; Ram, N.K. Performance analysis of an internal combustion engine operated on producer gas, in comparison with the performance of the natural gas and diesel engines. *Energy* **2013**, *63*, 317–333. [CrossRef]
22. Indrawan, N.; Thapa, S.; Bhoi, P.R.; Huhnke, R.L.; Kumar, A. Engine power generation and emission performance of syngas generated from low-density biomass. *Energy Convers. Manag.* **2017**, *148*, 593–603. [CrossRef]
23. La Villetta, M.; Costa, M.; Cirillo, D.; Massarotti, N.; Vanoli, L. Performance analysis of a biomass powered micro-cogeneration system based on gasification and syngas conversion in a reciprocating engine. *Energy Convers. Manag.* **2018**, *175*, 33–48. [CrossRef]
24. Chang, C.T.; Costa, M.; La Villetta, M.; Macaluso, A.; Piazzullo, D.; Vanoli, L. Thermo-economic analyses of a Taiwanese combined CHP system fuelled with syngas from rice husk gasification. *Energy* **2019**, *167*, 766–780. [CrossRef]
25. Cardoso, J.; Silva, V.; Eusébio, D. Techno-economic analysis of a biomass gasification power plant dealing with forestry residues blends for electricity production in Portugal. *J. Clean. Prod.* **2019**, *212*, 741–753. [CrossRef]
26. Brito, P.; Calado, L.; Garcia, B.; Alves, O.; Samanis, M. *A Critical Review on Acacia Gasification and Mix Acacia/Tires Co-Gasification, and Their Energy Assessments*; IPP: Portalegre, Portugal, 2017.
27. The Global Leader in Small-Scale Gasification. Available online: https://www.allpowerlabs.com (accessed on 15 March 2020).
28. Couto, N.; Silva, V.; Cardoso, J.; Rouboa, A. 2nd law analysis of Portuguese municipal solid waste gasification using CO2/air mixtures. *J. CO2 Util.* **2017**, *20*, 347–356. [CrossRef]
29. De Sales, C.A.V.B.; Maya, D.M.Y.; Lora, E.E.S.; Jaén, R.L.; Reyes, A.M.M.; González, A.M.; Andrade, R.V.; Martínez, J.D. Experimental study on biomass (eucalyptus spp.) gasification in a two-stage downdraft reactor by using mixtures of air, saturated steam and oxygen as gasifying agents. *Energy Convers. Manag.* **2017**, *145*, 314–323. [CrossRef]
30. Couto, N.; Silva, V.; Monteiro, E.; Rouboa, A. Exergy analysis of Portuguese municipal solid waste treatment via steam gasification. *Energy Convers. Manag.* **2017**, *134*, 235–246. [CrossRef]
31. CEPEA (Centro de Estudos Avançados em Economia Aplicada). *Setor Florestal*; CEPEA: São Paulo, Brazil, 2018; Volume 93.
32. Ferreira, A.L.D. *Culturas Energéticas: Produção De Biomassa E Bioenergia*; Universidade de Coimbra: Coimbra, Portugal, 2015.
33. Ferreira, S.; Monteiro, E.; Brito, P.; Vilarinho, C. Biomass resources in Portugal: Current status and prospects. *Renew. Sustain. Energy Rev.* **2017**, *78*, 1221–1235. [CrossRef]
34. Bilek, E.M.; Skog, K.E.; Fried, J.; Christensen, G. *Fuel to Burn: Economics of Converting Forest Thinnings to Energy Using BioMax in SOUTHERN Oregon*; U.S. Dept. of Agriculture, Forest Service, Forest Products Laboratory: Madison, WI, USA, 2005.
35. ERSE. *Proveitos Permitidos e Ajustamentos Para 2019 Das Empresas Reguladas do Setor Elétrico*; ERSE: Lisboa, Portugal, 2019.
36. ANEEL. *Relatório Pesquisa Iasc 2018 Brasil, Categorias e Distribuidoras*; ANEEL: Brasilia, Brazil, 2019.

37. Abrantes, J.M. *Avaliação Técnica e Económica da Aplicação de Sistemas Waste to Energy no Tratamento de Resíduos Urbanos em Aglomerados de Média e Pequena Dimensão*; Instituto Superior Técnico Lisboa: Lisboa, Portugal, 2016.
38. Ministerio das Finanças. *Guia Fiscal 2019*; Ministério das Finanças: Lisboa, Portugal, 2019.
39. Receita Federal do Brasil. *Capítulo XIII—IRPJ—Lucro Presumido 2017*; Receita Federal do Brasil: Brasilia, Braszil, 2017.
40. Lyons, W.C.; Plisga, G.J.; Michael, D. Lorenz Petroleum Economic Evaluation. In *Standard Handbook of Petroleum and Natural Gas Engineering*; ELSEVIER: Amsterdam, The Netherlands, 2016; pp. 7-1–7-55, ISBN 9780123838469.
41. Investopedia Corporate Finance & Accounting. Available online: https://www.investopedia.com/terms/c/cfat.asp (accessed on 4 June 2020).
42. Diário da República. *Decreto Regulamentar n.º 25/2009 de 14 de Setembro. Diário da República, 1.ª Série—N.º 178—11 de Setembro de 2015*; Diário da República: Lisboa, Portugal, 2009.
43. ANEEL Biomassa. *Atlas de Energia Elétrica no Brasil*; ANNEL: Brasilia, Braszil, 2005; Volume 5, pp. 77–92, ISBN 978-85-87491-10-7.
44. BBC. Los Países de América Latina Donde se Pagan Más y Menos Impuestos—BBC News Mundo. Available online: https://www.bbc.com/mundo/noticias-47572413 (accessed on 10 March 2020).
45. Cardoso, J.; Silva, V.; Eusébio, D.; Brito, P.; Tarelho, L. Improved numerical approaches to predict hydrodynamics in a pilot-scale bubbling fl uidized bed biomass reactor: A numerical study with experimental validation. *Energy Convers. Manag.* **2018**, *156*, 53–67. [CrossRef]
46. Silva, V.; Monteiro, E.; Couto, N.; Brito, P.; Rouboa, A. Analysis of syngas quality from Portuguese biomasses: An experimental and numerical study. *Energy Fuels* **2014**, *28*, 5766–5777. [CrossRef]
47. Cardoso, J.; Silva, V.; Eusébio, D.; Brito, P.; Hall, M.J.; Tarelho, L. Comparative scaling analysis of two different sized pilot-scale fluidized bed reactors operating with biomass substrates. *Energy* **2018**, *151*, 520–535. [CrossRef]
48. Homdoung, N.; Tippayawong, N.; Dussadee, N. Performance and emissions of a modified small engine operated on producer gas. *Energy Convers. Manag.* **2015**, *94*, 286–292. [CrossRef]
49. Glenting, C.; Jakobsen, N.F.B. *Converting Biomass to Energy. A Guide for Developers and Investors*; World Bank Group: Washington, DC, USA, 2017.
50. Vera, D.; Jurado, F.; Margaritis, N.K.; Grammelis, P. Experimental and economic study of a gasification plant fuelled with olive industry wastes. *Energy Sustain. Dev.* **2014**, *23*, 247–257. [CrossRef]
51. Rentizelas, A.; Karellas, S.; Kakaras, E.; Tatsiopoulos, I. Comparative techno-economic analysis of ORC and gasification for bioenergy applications. *Energy Convers. Manag.* **2009**, *50*, 674–681. [CrossRef]
52. Szabó, S.; Bódis, K.; Huld, T.; Moner-Girona, M. Energy solutions in rural Africa: Mapping electrification costs of distributed solar and diesel generation versus grid extension. *Environ. Res. Lett.* **2011**, *6*, 034002. [CrossRef]
53. Amirante, R.; Bruno, S.; Distaso, E.; La Scala, M.; Tamburrano, P. A biomass small-scale externally fired combined cycle plant for heat and power generation in rural communities. *Renew. Energy Focus* **2019**, *28*, 36–46. [CrossRef]
54. Mainhardt, H. *Funding Clean Energy Access for the Poor: Can the World Bank Meet the Challenge*; Bank Information Center Europe: Amsterdam, The Netherlands, 2017.
55. Zanchi, G.; Frieden, D.; Pucker, J.; Bird, D.N.; Buchholz, T.; Windhorst, K. Climate benefits from alternative energy uses of biomass plantations in Uganda. *Biomass Bioenergy* **2013**, *59*, 128–136. [CrossRef]

MDPI
St. Alban-Anlage 66
4052 Basel
Switzerland
Tel. +41 61 683 77 34
Fax +41 61 302 89 18
www.mdpi.com

Energies Editorial Office
E-mail: energies@mdpi.com
www.mdpi.com/journal/energies